Selected Essays
on *Enchantress from the Stars* and More

A Sampler from My Essay Collections

Sylvia Engdahl

* Ad *
Stellae

Eugene, Oregon
2024

ISBN: 979-8985853285

Contents

Contents

Preface

This book contains a few of the essays from my three volumes of collected essays, two of which are available only as ebooks. The printing cost of those collections would be so high as to make paperback editions prohibitively expensive. Yet there have been requests for them, and there are many people who would be interested in the background of my novels who for one reason or another don't want to read ebooks. Also, I would like to have a print edition of the most important essays available in libraries. So I have produced this short collection, keeping within the number of pages that will permit the lowest price for a print-on-demand paperback the distributors allow.

All but one of the essays included here are in *Reflections on Enchantress from the Stars and Other Essays* or *The Future of Being Human and Other Essays*. (There is only one from *From This Green Earth: Essays on Looking Outward* because that book has been shortened and is now available in paperback.) These essays do not need to be read in order; they are independent. I know some people are interested only in *Enchantress from the Stars*, which is quite different from my other novels (and the only one suitable for pre-teen children as well as older readers), but those included here concern ideas that are relevant to *Enchantress* even when they don't mention it. They shed light on how much I really believe of what my fiction implies about human nature and the future of humankind.

—*Sylvia Engdahl, May 2024*

Reflections on *Enchantress from the Stars*

This is a more detailed and formal presentation of things I have been saying at my website and elsewhere for many years. I hope that people interested in the book will read it, especially teachers who have discussed the story with young readers.

*

Authors are not supposed to tell readers how to interpret their books. Ordinarily it should be left to each reader to do so in his or her own way, and if the author fails to convey the intending meaning in the story itself, that is a flaw in the writing of it that no amount of explanation can remove. But in the case of *Enchantress from the Stars*, there has been confusion arising from its science fiction content that I feel needs clearing up.

Enchantress from the Stars is a book with more than one level and there's much in it that I believe applies to people of our own time in our own world—ideas about the different ways truth can be seen, about the power of faith, and about love. But as I have been saying since its first publication, it dismays me when readers assume it is a wholly allegorical story rather than one literally about relations between species that evolved on separate planets. That was something I didn't anticipate, and I don't quite see why, in an era when respect for all cultures is viewed as important, so few people see the harm in it. Usually it's okay for a book to be interpreted differently by different readers—but not when a common misinterpretation gives the impression that the author endorses a view of cultural relations on Earth that is generally considered not merely mistaken but "politically incorrect."

To assume that the premises of *Enchantress* apply to relations among groups of the same species is a false analogy, and it leads to the conclusion that I view cultural differences in a way that was rejected by anthropologists long ago. Since I came close to getting a master's degree in anthropology I don't like having it thought, as it is by some critics, that I'm either ignorant or intentionally promoting that view, even apart from the fact that it's not one that young readers should be encouraged to adopt.

It used to be thought that some cultures on Earth were "primitive" while our own culture was "advanced" in a more fundamental sense than its possession of modern technology. Today this idea is looked upon as obsolete and condescending. All independent cultures on this planet have been developing for the same length of time, although some have changed more than others. We are all members of the same human race, the same species. The different peoples in *Enchantress*, however, are of *different* species, some of which are biologically older than others and whose civilizations have existed for longer periods of time. The variance in their maturity is evolutionary, not merely cultural. Relations between them cannot be compared to relations among people with the same origin. Moreover, basic to the premise that interstellar contact would be detrimental to young species is the fact that the existence of more mature ones is unknown to them, a situation that cannot exist on any single world.

To be sure, the fundamental idea that it's wrong to treat others as subhuman and seize land that belongs to them does apply to Earth. But when readers carry the analogy further, the story seems to be saying that we should not offer any help to developing nations or to societies on our own world whose members are sick or starving, which I certainly didn't mean to imply. Elana's people hold that it would be harmful to give aid to less mature species because it would interfere with their evolution and prevent them from eventually making a unique contribution to the community of advanced civilizations. (Which is why I believe extraterrestrials will not contact us at our present stage, much as advocates of SETI hope they will). Extragenetic evolution, however, applies to a planetary civilization as a whole; it cannot be said that some groups of the same species are further evolved than others.

Some readers have felt that the Federation in the story is rather high-handed in labeling the inhabitants of some worlds "mature" while others are not, and this would be a valid criticism if they had not been evolving for different lengths of time. In actuality, there is nothing arbitrary about the threshold I envision. The more advanced technology and less inhumane customs of the mature peoples as compared to the "Younglings" are *consequences* of their species' age, not random characteristics by which they are subjectively judged. It is to be assumed that different cultures exist on all worlds, as they do on ours, though for sake of simplicity the story doesn't show that; yet the civilization of each world *as a whole* either has reached a level where it can meet other worlds' planetary civilizations as an equal, or it has not. This, of course, is not to say that all individuals of a given species are equally mature. In my novels only the agents of the Service, who are selected according to very high standards, are allowed contact with "Youngling" worlds, so the variations among members of mature civilizations are not mentioned. The level of a species, however, depends on the qualities of the majority of its people, which need not be possessed by all of them.

What defines that level? As I have said in the novels of my *Captain of Estel* trilogy, as well as by implication in *Enchantress*, I believe it is the widespread development of consciously-utilized psi (psychic) powers, especially telepathy. Not only would such powers lead to a greater degree of understanding and empathy than exists among the people of a world at our present stage of evolution, they would be essential to contact with extraterrestrials whose physical appearance would offer none of the clues on which communication has depended since the dawn of history. Without telepathic rapport the gulf between species would be too wide to cross, and hostility or an intent to exploit would be suspected where none existed. Moreover, people who lacked such capability could not function effectively in an interstellar society based on it; they would feel isolated and deficient no matter how much respect they were given.

By telepathy, of course, I do not mean "mind reading"—telepathy as I see it is two-way communication and is voluntary, at least at the unconscious level, on both sides. It is latent in all of us and has influenced history to a far greater extent than is imagined. The degree to which it can come under conscious control is unknown, and the use of it in my fiction does not pretend to be a realistic portrayal of a faculty beyond our present understanding. Undoubtedly it would not take the form of conversation as it has to be presented in writing; I suspect it would be entirely wordless. And a society in which it was common would not be as much like ours as the ones in the stories.

Whether any other psi powers ever approach the level described in my novels is an open question. I have intentionally exaggerated them not just for plot purposes, but to symbolize my belief that evolutionary advancement is not merely cultural but involves factors beyond our ability to truly imagine. I feel sure that we will ultimately develop conscious control of telepathy, but it's unlikely that future evolution will give us the ability to place our hands in fire without being burned, as Elana and the characters in my adult novels do. That is meant simply as an indication that evolving far beyond our present stage would involve developing capabilities that sociocultural change cannot produce.

There is another reason why I'm sorry that so few readers take the relationships between worlds in *Enchantress* seriously. One of my aims in writing it was to influence young people's attitude toward extraterrestrial aliens. In the movies and in prevalent UFO lore, aliens are generally portrayed as hostile and repellent. On the other hand, some people view extraterrestrials as benevolent "gods from outer space" who would either consider the problems of Earth evidence we are an innately deficient species and a danger to the galaxy, or would tell us how to solve those problems—as in fact some scientists hope they will if radio contact with them can be made. In my opinion none of these ideas are constructive. They don't encourage effort to solve our own problems, and what is worse, they foster a negative view of the wide universe from which Earth cannot remain isolated.

This is not an issue to dismiss as silly or inconsequential. It doesn't matter whether any aliens show up within the lifetime of young people living today (which, for the reasons given in my novels, I personally believe they won't). The view of our place in the universe absorbed by the young will be passed from generation to generation and will shape the future of our civilization. It may even affect the pace of our progress toward becoming a starfaring species, which I consider essential to our long-term survival. And if representatives of advanced extraterrestrial ones ever do appear, we surely don't want assumptions drawn from alien invasion stories to affect our reception of them.

For both these reasons, I have mixed feelings about the commonly-expressed idea that *Enchantress from the Stars* is "half fantasy and half science fiction." There is no fantasy in it, except in the sense that all science fiction contains material that is purely imaginary. Portions of it are told in the *literary style* of fantasy, which is something quite different from having elements of fantasy in the story. Insofar as the misconception attracts readers who enjoy fantasy more than typical science fiction, calling it a mixture is a good thing; yet it also leads to the assumption that none of the story is meant to be taken literally. All good fantasy has a level on which the outlook toward life embodied is, in the author's opinion, valid—but on the level of the story's action it is not intended as serious speculation. Even when it is satire on past or present human events, it does not attempt to say more about the future than that mistakes of the past should be avoided. The creator of a fantasy world does not expect readers to wonder whether a comparable world, or situation, really exists somewhere. It is taken for granted that it is imagined simply to highlight thoughts about the here and now.

Science fiction, on the other hand, usually does say something about the future, albeit not at all literally with regard to the details. For example, "space opera" about battles with aliens assumes and instills the idea that because war has been common among humans, war with extraterrestrials is to be expected. Such ideas may not reflect the author's actual views—though often they do—but they affect how readers feel about what lies ahead for humankind. This is particularly true of young readers. They don't necessarily believe the underlying premises of a story consciously, but they absorb them unconsciously and pass them on. If a story is set in the future, it will have an emotional impact on attitudes toward the future, and calling it fantasy will not lessen that impact. And I think that the extent to which this is recognized by classifiers matters.

I would like to believe that my readers' feelings about our place in the larger universe is influenced, to at least a small degree, by imagining what relationships with extraterrestrial species might involve. So it disturbs me when they are led to think that the different peoples in the novel are merely different cultures of our own species in disguise. It is true that in one respect my portrayal of them is indeed a mere reflection of human cultures, since it is based on human mythology—on the comic-book image of space explorers prevalent in the twentieth century as much as on the myth of dragon-slayers derived from fairy tales. This does confuse the issue somewhat, especially since to my surprise many adult readers thought the invaders in the story were stereotyped rather than intentionally depicted in terms of a modern myth comparable to the older one.

Actually, the Imperials were no more meant to be realistic than the medieval characters. Moreover, they are an anachronism; a civilization advanced enough to build starships would not behave as our ancestors did. Even today, no one in a position to form space policy would consider colonizing a planet that has indigenous inhabitants. Yet the basic idea that humans, and presumably extraterrestrial civilizations, do advance—that the invaders were immature rather than collectively evil and could be expected to outgrow their aggressiveness—is, I trust, powerful enough to override such incongruities. It is this concept that should be taken literally, not the details that form the story's plot.

And it is this concept, more even than the novel's premise about relationships between our world and others, that I hope young people will absorb. Few readers have any grasp of the time scales involved in human progress. They are thus apt to think no progress is made, despite the fact that men of the past viewed war as glorious and desirable, and slavery was considered normal even in this country less than two hundred years ago. Obvious though it is that most of us have become a great deal less barbarous than the people of ancient times, all too often today's problems are seen as an indication that human behavior will never be any better—yet lack of change within a few generations means nothing in view of the centuries yet to come. Elana's people judge progress from the perspective of experience with many planetary civilizations over a long period of time. I hope imagining such a perspective will help readers realize that neither our own future actions nor those of any aliens we may meet can be predicted on the basis of our present stage of evolution.

I believe that we have an exciting future ahead among worlds beyond Earth, whether we meet any extraterrestrials or not. And it is not too soon to start caring how that future is envisioned, if only for the effect on the outlook of people today. That is why I write science fiction, and why the way my novels are interpreted matters to me.

Perspective on the Future:
The Quest of Space Age Young People

This essay was adapted from a speech I made to the Washington State Association of School Librarians in March, 1972 and first published in School Media Quarterly, *Fall 1972. It was reprinted in* Only Connect: Readings on Children's Literature, *S. Egoff, G.T. Stubbs & L.F. Ashley, eds., Oxford University Press, 1980.*

*

Those of us who work with literature for youth have many things in common, whether we are writers, librarians, or teachers—and I believe that one of them is a very strong and basic interest in the future. I have been fascinated by ideas about the future, and particularly about space exploration, since I myself was in my teens; all my novels have been focused on it. While educators may not have such specific enthusiasm for the subject of the distant future, all are deeply concerned with preparing young people to live in the world of tomorrow. None of us can predict just what that world is going to be like, but I think there is much we can do to equip the next generation to cope with whatever tomorrow brings.

I suppose every author is asked how he or she came to choose subjects, but I think the question is raised more frequently with authors of science fiction than with others. People are always curious about why anyone would choose to write about imaginary things instead of the things we know. Each author has her own reasons, and mine are not really typical; perhaps an explanation of them will make clear why I feel that stories that deal with the future are important, and are of interest even to those for whom neither science fiction nor science itself has any special appeal.

First of all, I should mention that my books are more for a general audience than for science fiction fans. Although I think science fiction fans will enjoy them, I aim them principally toward people who normally do not read science fiction, and I avoid using esoteric terminology that only established fans can understand. Actually I am not what one would call a fan myself, at least not in the sense of keeping up with the adult science fiction genre. I use the science fiction form simply because my ideas about humankind's place in the universe can best be expressed in the context of future or hypothetical worlds.

This is not to say that my books are wholly allegorical. I have been rather dismayed to find that some people interpret them that way, because although there is indeed a good deal of allegory in them, they also have a literal level. For instance, what is said in *Enchantress from the Stars* and *The Far Side of Evil* about how a truly mature civilization would view peoples of lesser advancement is meant to be taken literally; scientists are beginning to ask why, if civilizations more advanced than ours do exist in other solar systems, they haven't contacted us, and that is my answer as to why.

Of course, one of my main reasons for writing science fiction is that I believe very strongly in the importance of space exploration to the survival of our species. I have held this belief since the days when all space travel was considered fantastic, and indeed I developed the theory of the "Critical Stage," on which my book *The Far Side of Evil* is based, in unpublished work that I did before the first artificial satellite was launched. I am entirely serious about the choice between expansion into space and human self-destruction being a normal and inevitable stage of evolution; the fact that when I came to write the book, our establishment of a space program had made it impossible for the story's setting to be Earth, as it was in my initial draft, was to me the most encouraging sign of our era. In the early fifties I had been afraid that the Space Age would not begin soon enough. [In the 30 years since this was published, the stalling of the space effort has shown more clearly than ever the need for fiction to inspire its progress.]

But apart from my commitment to the cause of space exploration, I think there is good reason to set stories in the future when writing for teenagers. Today's young people identify with the future. Many of them find it a more pertinent concept than that of the past. If we are going to make any generalization about the human condition, any convincing statement that evolution is a continuous process in which the now that seems all-important to them is only a small link, we stand a better chance of communicating when we speak of the future than when we describe past ages that—however mistakenly—the young have dismissed as dead and irrelevant. Teenagers are far more serious-minded than they used to be, yet they don't consider anything worth serious attention unless they see its relationship to problems they have experienced or can envision.

This has become more and more evident during the past few years. It so happened that I began writing in a period when young people's involvement with matters once thought too deep for them was increasing. I was not at all sure that there would be a place for the kind of novels I wanted to write, because they were too optimistic to fit the gloomy mold of contemporary adult fiction, yet too philosophical, I thought, to be published as teen fiction. Fortunately I directed them to young people anyway, and quite a few seem to like them. I don't think this would be the case were it not that the boys and girls now growing up are more mature in their interests than those of former generations.

It is apparent today that the young people of our time are searching desperately for something that they are not getting in the course of a standard education. They are searching in all directions: some through political activism; some through "dropping out"; some through renewed interest in religion in both traditional and novel forms, or even in the occult; and all too many through drugs or violence. Misguided though some of these attempts may be, I feel that they all reflect a genuine and growing concern on the part of our youth for a broader view of the universe than our present society offers them. Some can find meaning in the values of their elders; others cannot. There would seem to be a wide gulf between the two attitudes. There is a great deal of talk about polarization. Yet underneath, whatever their immediate and conscious goals, I believe that all young people are seeking the same thing: they are seeking a perspective on the future.

The need for such perspective is not new. It is a basic and universal human characteristic. What is different now is that the perspective inherent in the culture passed automatically from one generation to the next is no longer enough. Perspective implies a framework, a firm base from which to look ahead, and in this age of rapid change the old framework is not firm. Many of its components are still true and sound, but it has become so complex that as a whole it must necessarily invite question, if only because of the contradictions it contains. Scarcely anyone today is so naive as to suppose that all aspects of our current outlook are valid. There is much controversy, however, as to which are valid and which are not, and among free people the controversy will continue, for we live in an era when our civilization's outlook is constantly shifting and expanding.

Whether this is occurring because—as I believe—the time of our first steps beyond our native planet is the most crucial period in human history, or whether its basic cause is something else, the fact remains that it is happening. It is a confusing time for all of us, but especially for our young people, the members of the first Space Age generations, who are so aware of change and of the need for change that they can find nothing solid to hold to. They haven't the background to know that problems have been solved in the past, that present and future problems will in turn be solved, that the existence of problems is not in itself grounds for bitterness. They hear their disillusioned elders speak of the future with despair and they have no basis for disbelief. Yet instinctively, they do disbelieve—and I wonder if this, as much as the world's obvious lack of perfection, may not be why they find it so hard to believe any-thing else their elders tell them. They cannot accept the now-fashionable notion that the universe is patternless and absurd; they are looking for answers. Inside, they know that those answers must exist.

Young people cannot be blamed for thinking the answers are simple. Earlier generations have thought the same. But nowadays one's faith in a simple answer cannot survive very long; what Space Age generations need is awareness that one must not expect simple answers, and that humanity's progress toward solutions is a long, slow process that extends not merely over years, but over centuries. Knowledge of past history alone does not give such awareness because most of today's teenagers just don't care about the past. Significance, to them, lies not in what has been, but in what is to come. I believe that only by pointing out relationships between past, present, and future can we help them to gain the perspective that is the true object of their search.

One might wonder how I can consider this need for

perspective so fundamental when for years, psychologists have been saying that people's basic need is for security. Yet I think our young people are showing over and over again that they do not want security, at least not security as it has commonly been defined. A great deal of effort has been devoted to making them secure, yet many turn their backs and deliberately seek out something dangerous to do. The security they need cannot come from outside; it must come from within, from experiences through which each person proves that he or she is capable of handling the stresses of an indisputably insecure world. But no one can handle a situation in which he sees no pattern, no meaning. There can be no security without direction. Thus a perspective on the future is implicit in the very concept of inner security.

One's view of the future is, of course, a highly personal thing. Our beliefs can differ greatly as to the direction we are going, or ought to go. In my books I naturally present my own opinions, and I don't expect all readers to agree with them. But I hope that even those who do not agree will gain something by being encouraged to develop their private thoughts about the topics I deal with. I hope that they will be convinced that we are going somewhere, and that this will help to counter the all-too-prevalent feeling that human evolution is over and done with. It is this, more than anything else, that I try to put across: the idea that there is continuity to history, that progress—however slow—does occur, and that whatever happens to us on this planet is part of some overall pattern that encompasses the entire universe. We are not in a position to see the pattern. We can only make guesses about it, and many of those guesses are bound to be wrong. Still, I do not believe that guessing, either in fantasy or in serious speculation, is a futile task; for when we ignore the issue, we are apt to forget that the pattern exists whether we see it or not. That, I think, is the root of many young people's turmoil. They have no conviction that there is any pattern.

A common reaction to the space flights so far undertaken seems to be that we had better appreciate Earth because it's the only good planet there is. It is quite true that it is the only one in this solar system that is suitable for us to live on at present, and that those of this system are the only ones we have any immediate prospect of reaching. But the attitude that no other planet is worth anything strikes me as a new form of provincialism. Our solar system is merely a small part of a vast universe that contains billions upon billions of stars. People sometimes ask me if I really believe that there are habitable planets circling those other stars; the answer is that I do, and that most scientists now do also. Not everyone seems to realize this; several acquaintances told me rather shamefacedly that they themselves thought that there is life in other solar systems, although they were sure that scientists would laugh at them. As a result, I wrote a nonfiction book [*The Planet-Girded Suns*, 1974; updated edition published in 2012] that I hoped would explain to young people not only what modern scientists did believe, but what many philosophers of past ages believed about an infinity of worlds. The idea is not new, and it has not been confined to science fiction. Giordano Bruno was burned at the stake in the year 1600 for holding it.

Of course, I do not believe that the inhabitants of other solar systems are as much like us in the physical and cultural sense as I have depicted them in my novels. Most serious science fiction does not make them so similar, and I think that many potential readers are thereby turned away. They are put off by the weird element inherent in any attempt to imagine what sentient species other than ours would be like. I feel that this is distracting. Since we don't know what they are like and my aim is to show essentially identical spiritual qualities, it seems to me best to portray them in our terms, just as I have to make them speak in our language. Also, in *Enchantress from the Stars*, I wanted to leave open the question of which, if any, of the people were from Earth. Only in that way could I make my point about various levels of advancement.

This point, which is further developed in *The Far Side of Evil,* concerns evolutionary advancement, not mere cultural advancement. My intent was to comment upon relationships between eras of history, and between peoples at different stages of evolution, not relationships between societies here on Earth. We of Earth, whatever our nationality or our color, are all members of the same human race. We are one people, one species. Someday, generations hence, we may encounter other sentient species. It is not too soon for us to begin thinking about our identity as a people, our place in a universe inhabited by many; the young are better aware of that than most adults.

To those who do not believe that there will ever be contact between the stars, I would like to suggest that as far as contemporary youth's perspective is concerned, it makes no difference whether there is or not. The mere idea is, in itself, of consequence. I am troubled by science fiction's usual portrayal of advanced aliens either as hostile, or as presumptuous meddlers who take it upon themselves to interfere with the evolutionary process. The dangers of the first attitude are obvious; those of the second are perhaps less so. Maybe the whole issue seems remote and insignificant when we have so much else to worry about. Yet if young people acquire the idea that some extrasolar civilization could solve our problems for us if its starships happened to come here, or that it would consider our failings evidence that our whole human race is wicked instead of merely immature, will that not add to their already-great sense of futility? Will it not interfere with whatever perspective on human history they have managed to absorb? I think it will; and furthermore, whether there really are any alien civilizations is immaterial. Science fiction may be fantasy, but that young people like it and are affected by it is fact. It is also a fact that the Voyager probe launched by NASA carried a plaque designed to communicate its origin to any intelligent beings who recover it after it passes out of our solar system. It may be that no aliens will ever see that plaque, but our children saw it on television; their attitude toward its hypothetical viewers is bound to influence their attitude toward our own civilization.

Their view of civilization is already confused and inconsistent enough. On one hand, many believe that only scientific knowledge is factual, and that advancement is merely a matter of inventions and technical skill. On the other, during the past few years some people, especially the young, have come to distrust science, to blame it for our problems and even to question the value of technological advance—which, I believe, is the greatest distortion of perspective I have yet seen. Today, in their quest for meaning, young people are challenging the materialistic outlook many scientists have held in the past—and rightly so. At the same time, however, some of them are rejecting not only inadequate theories, but the whole idea of scientific progress. They seem to feel that in so doing they are defending spiritual values against some implacable enemy. They imagine that they seek a wider truth. Yet actually this viewpoint is equally narrow and in fact self-contradictory, for truth is precisely what science seeks, and has always sought from its very beginnings. There has never been any conflict between the real scientific attitude and spiritual values, where there appears to be; the trouble is with the particular theory involved and not with science as such. Truth is truth; science is simply the name given to the part we have attempted to organize and verify.

I think the current misunderstanding is the result of our tendency since the late 19th century to compartmentalize science, to separate it from the rest of life in the same way that some people separate religion. There was a time when the major scientific thought of an era could be understood by every educated person; but for many years now specialization has been necessary, and this has led to an unfortunate conception of what science is. Non-scientists have gotten the idea that it is some kind of esoteric cult that stands apart from other human endeavors, while both they and the scientists themselves have felt that its realms have been charted and need only to be conquered. When young people observe that there are things worth investigating outside these realms, and that some of our current scientific theories are questionable, it often doesn't occur to them that the answer lies not in abandoning science but in expanding it: refuting its dogmatic portions as dogma has been refuted countless times in the past. This, perhaps, is why some of them are turning in desperation to supernaturalism, astrology, and the like. Yet science is distinguished from superstition not by the subject matter with which it deals, but by the maturity of its explanations; it is distinguished from philosophy not by content, but by the availability of data to which objective scientific methods can be applied. All the phenomena now dealt with by science were once explained by superstition and, as an intermediate step, all our sciences were once divisions of philosophy. For that matter, there are advanced theories in all fields that are philosophic in that they are not yet subject to empirical proof. Because nowadays the people who hold such theories are called scientists and not philosophers, we get the impression that the theories are authoritative; but actually some are no more so than theories of the Middle Ages that have been disproven.

The point to be made is that this process of progression is by no means finished or complete. There is no area of truth that is outside the province of science in principle, though there are many that science lacks the practical means to investigate at its present stage of development. It is thus a great mistake to identify science with materialism, and to assume that it inherently deals only with the material aspects of the universe, when the fact is merely that these aspects can be more readily studied than other aspects that we are just beginning to rescue from the realms of the "supernatural." There is no such thing as the supernatural, since "natural," by definition, includes all aspects of reality. But too many of us have shut out parts of reality. We have discarded not only superstition, but also the areas with which superstition presently deals, forgetting that the superstition of today is merely an immature explanation of the science of tomorrow. We have failed to recognize that there are natural laws that cannot be explained in terms of the ones we know because they are, in themselves, equally basic.

Worse, our society has tended to assume that there is a firm line between science and religion. It has outgrown trust in superstition, and many have identified faith with superstition, discarding that also. Yet the fact that the physical aspects of natural law are the most readily analyzed does not mean that there isn't a spiritual reality that is just as real, just as much a part of the universe, as the material reality that science has so far studied objectively. I don't wonder that young people have difficulty in viewing the world with perspective when they have been led to feel that it is necessary to reject one or the other. The young today sense that moral and spiritual values are important, though they will not accept dogma in religion any more than in any other field, and it is understandably hard for them to reconcile their innate idealism with a science that is seemingly opposed.

To me, science itself can never be opposed to truth in any form whatsoever, no matter how many specific theories may be mistaken, and no matter how dogmatic certain scientists may be in support of their own era's beliefs. This is how I have viewed it in *Enchantress from the Stars*, and I think one of the book's appeals for young people is that it does take seriously certain things outside the traditional bounds of science, such as extrasensory perception, without putting a materialistic inter- pretation on them. I hope readers notice that nowhere have I suggested that advanced peoples, in progressing beyond a materialistic orientation, would give up any of their technology; because I feel strongly that as they matured, they would improve their technology and learn to put it to better use.

I am convinced, therefore, that the solution to future problems lies not in de-emphasizing science, but in advancing it, as well as in an outlook that recognizes that the science of any given age is imperfect and incomplete. For instance, I believe that while there is much that can and should be done now to slow the rate of population growth, the only permanent answer to overpopulation is the colonization of new worlds. I have been asked how I can approve of our colonizing planets in other solar systems if other sentient species exist. Certainly I don't think we should colonize planets that are already occupied; I trust my books make that very clear. What I do think is that there are many worlds on which no intelligent life has evolved that can be made livable by advanced technology, and that in the normal course of a sentient species' evolution, it expands and utilizes such worlds. There is nothing less natural in that than in our ancestors building the ships and other equipment needed to colonize America. Pioneering is a basic human activity; that's the comparison I tried to draw in *Journey Between Worlds*.

This question of what is natural for us seems to need a good deal of examination right now. There is a feeling prevalent today, particularly among young people, that we ought to get "back to nature." Insofar as this means preserving and enjoying the beauties of our world, it is a good thing. But those who say that we as a species should live in a more "natural" way are, I think, overlooking what "natural" means as applied to human beings. It is the nature of animal species to remain the same from generation to generation, evolving only as adaptation to physical environment may demand. It is the nature of our own species, however—and of whatever other sapient races may inhabit this universe—to learn, to change, and to progress. There is no point at which it is "natural" to stop, for to cease changing is contrary to the mental instincts that are uniquely human. If it were not so, all learning, from the discovery of fire to the conquest of disease, would be unnatural, and I don't think anyone believes that—least of all the young, who are more eager for change than their elders. It is the nature of humans to solve problems. It is the nature of humans to grope continuously toward an understanding of truth. There may be disagreement as to means, disagreement as to what is true and what is not, but never on the principle that to search for truth is an inherent attribute of humankind.

In my novels *This Star Shall Abide* and its sequel *Beyond the Tomorrow Mountains* [the first two volumes of the Children of the Star trilogy], I said quite a bit about the search for truth, from both the scientific and the religious standpoints; and I also tried to say something about the importance of faith. Yet the people of these stories are stranded in a desperate situation where only advanced technology, and an eventual major advance in scientific theory, can prevent their extinction. To achieve this advance, they are dependent on the kind of creative inspiration that has underlain all human progress since the beginning of time. Their religion is central to their culture, and it is in no way a materialistic religion; but the hope it offers them can be fulfilled only through faith in the ultimate success of their scientific research.

I wrote a description of these two books for Atheneum in which I defined science as the portion of truth that no longer demands faith for acceptance. That's the way I look at science: it is part of a larger truth. I believe that if we can give young people that sort of attitude toward it—if they can be helped to view its failure to provide all the answers overnight with neither hostility nor despair, but with the willingness to keep on searching—we will go a long way toward building their perspective on the future. And I believe that it is such perspective, more than anything else, that will fit them to take their place in tomorrow's world.

Faith as the Focus of *Children of the Star*

The first part of this essay was originally a pamphlet distributed by the publisher to librarians at the time This Star Shall Abide *first appeared. I have written the rest, portions of which are in the FAQ at my website, to address questions raised by* The Doors of the Universe. *Be aware that it contains major spoilers.*

*

It is a strange experience for a writer to find more themes inherent in a story than were originally meant to be there. This has happened to me before, but never to the same extent as in *This Star Shall Abide,* and later in the complete trilogy *Children of the Star.*

The ideas for my three previous novels, as well as for this one, came to me some in the late 1950s. I was not free to write the books then, nor would they have been thought timely. When I did write them, it was largely a matter of finding ways to express what I had long wished to say. But the story that became *This Star Shall Abide* (known in the UK as *Heritage of the Star)* underwent more change during its development than did the others. I always knew that the subject of space exploration and of humanity's place in the universe would be timely someday; it has been my prime concern for as long as I can remember. That the issue of youthful heresy would become central in our society was less obvious to me. And certainly I did not foresee that the concept of technological innovation as essential to human progress would ever be questioned, as it is questioned by some today, thereby giving the story's setting a thematic importance of its own. Thus what began as a relatively simple story grew into one so involved that its telling demanded not one volume, but two, of which *This Star Shall Abide* was the first.

This Star Shall Abide is about a boy who rebels (justifiably) against the religious and secular authority of his world; who is convicted of heresy, which in the eyes of his people is a serious crime; who refuses either to recant under pressure or to sell out for personal gain—and who is then confronted with proof that most of his beliefs are mistaken. The people's religion is not mere superstition. Their social system, although condonable only as the lesser of two evils, is not corrupt. Their world is not as it should be; it is not as it will become; but there are valid reasons—unique ones not comparable to any that could arise in our world—why no immediate change is possible, and the goal of change must be pursued by constructive rather than destructive means. Noren, in the story, has the courage to acknowledge this, whereupon he discovers that the heretic's path leads to rewards and burdens beyond any he has ever imagined.

Something that seems false or foolish to a person whose knowledge is limited may turn out to be true ... it is possible that outward appearances are misleading and that more complete information would give him a different viewpoint ... it is right to question orthodoxy; truth must not be accepted merely on the grounds of authority or tradition; yet on the other hand, it must not be rejected merely because it *is* orthodox, since to do so is equally dogmatic. These, among other things, were the intended themes of the story. Initially I considered their religious aspects alone. In taking up the material after a lapse of years, however, I realized that they had acquired wider significance. For many young people heresy had become a way of life: not just the rejection of traditional religious symbols, but outright repudiation of society's established values. I realized that contemporary readers would find Noren's defiance of secular authority even more relevant than his scorn of a seemingly unreasonable faith.

While this fact added pertinence to the story, at the

same time it created complications. If the social system that
Noren defies were a good one, the book would not say
anything to young people who see only the bad features of
ours; they would dismiss it as a defense of the system itself.
The science fiction form was thus of great advantage to me
in that I could devise a system totally unlike ours and
wholly indefensible apart from the singular conditions under
which it exists. But science fiction, to me, is more than
allegory; science fiction must deal not only with what is true
for the people of our planet, in our time, but with what I
believe about the universe and about underlying truths
applicable to whatever sentient races may inhabit it. Having
once established a situation in which a human race is faced
by disaster so great as to justify an admittedly evil system as
the sole alternative to extinction—something I am
convinced does not occur on this planet, nor indeed
anywhere, in the normal course of evolution—I found that I
could not simply stop there. To do so would be to leave
readers wondering how such a situation could be reconciled
with the optimism I'd expressed in the past.

Moreover, the situation could not in one volume be
made entirely clear to readers, for Noren does not learn the
full truth about his world until the story's climax. He cannot
see why his people have reverted to ways more primitive
than those of their ancestors; he cannot understand that their
inability to progress is the inevitable result of their being
thrust into an alien environment with resources so limited as
to make technological advance virtually impossible. Nor can
he truly comprehend the hope their religion offers them. He
lacks the background to grasp the real nature of the means
through which that religion's promises are to be fulfilled.

Hence the second volume, *Beyond the Tomorrow
Mountains*. In *This Star Shall Abide* Noren seeks and
eventually finds, but what he finds is not the ultimate truth
he assumes it to be. Thus the time comes when he seeks

answers that can be found neither in the awesome City of his world nor, for that matter, in any other. It eventually occurs to him to ask *why* his people suffered a tragic setback that no human wisdom could have prevented, and why anyone who knows the facts should be sure that the setback can be overcome. He then challenges the system anew, unaware that those who have won the right to share its secrets are no less dependent on faith than those who give a literal interpretation to the symbols.

For in the last analysis, faith—or in other words, a positive view of the universe—is essential to all progress. Science is necessary, but one cannot rely solely upon science, which by definition concerns only that portion of truth that no longer demands faith for acceptance. This is something Noren has yet to discover at the conclusion of *This Star Shall Abide,* and something that our own society too seldom recognizes. It is as basic to this story as to my earlier ones, a part of the universal pattern for which today's young people, like Noren, are desperately searching. I hope that these books offer them encouragement in that search.

*

I had no plans to write a third volume. I felt that to let Noren discover a way to change his world would weaken the emphasis on faith with which the second one ends. In real life faith in an eventual solution to a major problem is rarely vindicated within a few years, or even within a lifetime. Certainly the problems in our own society cannot be solved quickly—the chief trouble I saw in young people's outlook was that they expected that they could be eliminated if only people would try hard enough. A main point of the story was that one must keep working toward a solution even when there is no hope of finding it in the foreseeable future.

But six years after *Beyond the Tomorrow Mountains* was published I learned to my dismay that the people of

Noren's planet might have been enabled to survive without the drastic system imposed by the Scholars. When I wrote the story, I myself was unaware of any other way they could have been saved. I believed that there was no alternative to what the Scholars did, for of course, I would not have sanctioned their actions on any lesser basis than my conviction that the extinction of their human race would have been worse. I knew nothing about genetics, and when I began to research it for another project, I was appalled, for I feared that informed readers would assume that I had ignored it for plot reasons and had knowingly justified the social evils in the story on false grounds. In any case, the story had to be updated.

And so I wrote *The Doors of the Universe*. The first problem I faced was explaining why the Scholars' knowledge had been incomplete. It was hard to think of a good reason, but at that time there was a lot of public opposition to genetic engineering and many people believed it should never be used on humans, so though I didn't really think a civilization as far advanced as that of the Six Worlds would ban it, I felt the premise that it did would hold water. I trust it still does, now that the public's negative reaction to genetic technology is less strong.

I was also concerned because I felt the new book could not possibly be considered a "children's book" when it not only dealt with Noren's adult life, but would necessarily include discussions of sex and even sexual encounters. Yet it would have to be issued by publisher's children's department because the first two books in the story had been. (At that time no Young Adult category existed.) I was quite surprised when my editor agreed, but found that she was already looking ahead to the publication of books for mature teens. Neither *Beyond the Tomorrow Mountains* nor *The Doors of the Universe* is of interest to readers below high school age, and when the trilogy was later published in a single volume by a different publisher it was issued as adult science fiction, which after all is what most mature teens read.

In developing the plot of the third book, I soon that realized that it depended even more on faith than the previous one. At the end of *Beyond the Tomorrow Mountains* Noren discovers to his surprise that underneath, despite his disillusionment about the failure of the scientific work that would enable his people to survive without the restrictive social system that has been imposed on them, he does have faith in that work's ultimate success. He thus gains understanding of why the Scholars find their religion meaningful even when they know its symbolic expressions aren't true in a literal sense. But this faith cannot last. In *The Doors of the Universe,* his increasing expertise as a scientist shows him that the other scholars' goal of synthesizing metal is unrealistic and that they are counting on him for a breakthrough he can never achieve. Even worse, when he learns that genetic engineering could make survival possible, he knows that it will mean the final abandonment of their hope of fulfilling the Prophecy—a hope on which everything he personally values depends.

Noren alone is willing to abandon it; the others are not, for if they did so, knowing the Prophecy can't come true, they could no longer serve as priests without hypocrisy. Only faith in his ability to succeed in the genetic research— and to somehow gain their support, and later, that of the common people who will resist change—can enable him to do what must be done in the face of the sacrifices it will demand. Yet how can he proceed when he's aware that even if he manages to ensure the survival of many generations, without fulfillment of the Prophecy his human race will eventually be doomed?

He'd once assumed that the faith he found earlier would be permanent. But, he sees suddenly, "That had been faith in which he'd had *no choice*. No choice but to die, anyway, as they'd all have died in the mountains if his subconscious faith had not sustained them. That was one kind, a necessary kind: simply to go on because there was nothing else to do. But it demanded no real action. . . . All at once he perceived what an act of faith involved. There had to be choice in it, a decision that might go either way; one must *choose* a road that might lead nowhere."

Noren chooses to strive for genetic change without imagining how he can persuade the world to accept it, without conscious faith that it won't prove futile in the long run. One step at a time, through many struggles, he becomes ready to offer it to the people despite opposition from his fellow Scholars. Yet he can do that only as a priest, in violation of his knowledge that even when seen as symbolic, the traditional religious phrases have become hollow. Or have they? Acknowledging that "the Prophecy is a metaphor, not a blueprint," he takes a step more drastic than anything he has foreseen. I myself had no idea how Noren could win the people over until I wrote the last chapter of the book; that was why it took me over a year to finish it. At the point where he finally sees, where "inspiration, when it came, was a flash of light," that was true of me also.

I have found that not all readers realize that the religious aspects of the story are meant to be taken seriously. I have received mail about *This Star Shall Abide* from people of several different religions, including members of the clergy, who admire it; and it won a Christopher award for "affirmation of the highest values of the human spirit" from a Catholic organization (though I am not Catholic). On the other hand, a few atheists have interpreted it as an endorsement of their views. Surprisingly, however, many people think the trilogy is about "a false religion" with no application at all to real ones.

Individual readers may, of course, consider any religion

other than their own "false." But when people speak of a "false religion" in the context of this trilogy, or of any science fiction, they usually mean something more like "fake religion." Does the mere fact that its central symbol, the Mother Star, was purposely chosen by the First Scholar make it fake? Or the fact that the Star doesn't really have supernatural power and that what people say about it isn't literally true? Those are questions readers will have to answer for themselves after reading the second and third novels. Personally I believe that many religious ideas are metaphors that cannot be taken as literal fact, but nevertheless express concepts that we have no better way of expressing. As Noren discovers, they are symbols of "the unknowable." Metaphor often conveys truth, and is in fact the only way of conveying truth beyond our rational understanding. To me, it is what the metaphor stands for that's important.

For the sake of readers who didn't grasp that Noren's people aren't members of our own human race, I should make plain that they are of an entirely different origin. I had assumed that by describing the home civilization in the story as "the Six Worlds" and stating that there were six very similar to each other in that solar system, it would be clear that it wasn't ours; but the reviewers of *This Star Shall Abide*, some of whom weren't knowledgeable about astronomy, didn't all see that. It's important because I didn't want to imply that the traditional religions of Earth will be forgotten by future colonists. Not only might that offend some readers, but it wouldn't fit the story; a new religion developed by our descendants wouldn't have the same characteristics as the one of Noren's planet.

There are also other reasons, apart from the fact that I wouldn't want to postulate our world's destruction. All my fiction takes place in the same "universe"—ours, as I imagine it may someday be, though of course the details aren't meant as predictions. To picture the events in *Children of the Star* as happening to our descendants would settle the question of whether Elana's people in *Enchantress from the Stars* are our descendants or visitors to our ancestors (or neither), which is intended to be an open one. Moreover, it wouldn't be consistent with my later novels, which are about the future of Earth and its many colonies in other solar systems.

Readers may wonder why I brought the Service from the Elana books into the third book of the trilogy when the first two originated as an entirely separate series. It was because I'd done such a good job of establishing that the planet hadn't enough metal to restore technology that no solution other than subtle aid was possible. If the Founders had known what the Service knew about recovering trace metals (which incidentally, is something that has already been done with genetically engineered bacteria here on Earth) they would have used that knowledge in the first place. So to achieve a happy ending, aliens had to help without revealing themselves. Furthermore, having established the presence of an alien artifact in *Beyond the Tomorrow Mountains*, I had to follow it up. And if there were going to be aliens with the same policy as the Service in the Elana books, making them a separate group would have looked too repetitious.

Incidentally, the finding of that artifact, which one reviewer called "deus ex machina," was not a matter of my being unable to think of a less coincidental way of getting my characters out of a jam. It was meant to imply that coincidences sometimes occur that aren't due to mere chance—cases of the mysterious phenomenon known as synchronicity. Or you can call it divine providence. That was the point, after all. How could the book have ended as it does if the rescue had not been unforeseeable through reason? This was one of the remarkable ways in which the first two novels turned out to contain unplanned preparation for the concluding one, because the arrival of the Service depended on the artifact's discovery, and that, too, needed to be not merely unforeseen but unforeseeable; otherwise the actions of the Scholars would have been unjustified.

Some readers have assumed that the unexpectedly-happy ending in the Epilogue was necessitated by the books having been originally published as Young Adult instead of adult novels. This wasn't the case; I wouldn't write even an adult novel with a tragic outcome for a whole civilization. An open ending, yes, as I did when I planned to conclude with *Beyond the Tomorrow Mountains,* but not outright disaster. And unlike some readers I've talked to, I believe the permanent loss of high technology would be a disaster that would lead ultimately to the species' extinction, even if its survival had been prolonged. I don't agree with the view that a primitive low-tech lifestyle can be indefinitely sustained, and in any case I believe that no species can last forever if confined to a single planet.

Moreover, mere survival and abandonment of the caste system would not raise the society to the level once attained by its people's ancestors. One theme of the story is that the loss of technology leads to loss of everything else that goes with advancement, including attitudes toward gender equality—a point missed by reviewers who complained about sexism in *This Star Shall Abide.* More is preserved in the City than machines. Women outside necessarily devote most of their time to childrearing since large families are needed to increase the planet's population, so of course the villagers, unlike the City dwellers, have sexist attitudes. I no more advocate this than I advocate their custom of lynching heretics! But that is what would happen in a low-tech society forced to revert to backward ways; survival of their species was not the only thing at stake in the happy ending.

Nevertheless, the main consideration was that restoration of interstellar travel was essential to the long-term survival of Noren's people, just as the development of it prior to the nova proved essential; and it is equally necessary to ours. The story is thus highly relevant to our own world, in view of scientists' current belief that faster-than-light travel will never become possible—a parallel that I intended to be seen in *Beyond the Tomorrow Mountains* with respect to the goal of synthesizing metal. We too need faith that there will be breakthroughs we cannot foresee.

Does it matter whether a human race survives indefinitely? My characters take it for granted that it does. Extinction of our species (or any comparable one) would in my opinion be an unmitigated evil. But recently I found to my surprise that some people don't think so. In reply to an essay in which I mentioned that almost everyone has a deep, instinctive feeling that our species will continue to exist when we ourselves are gone, one commentator wrote, "I guess you'll have to include me out . . . around the same time I was fascinated with dinosaurs, the idea that homo sapiens could also go extinct didn't seem at all ridiculous to me." Presumably it still doesn't bother him.

We do not know, of course, why our human race or any other exists in the first place, so we have no basis on which to judge whether its survival matters in the long-term scheme of things. Yet to say that it doesn't removes the foundation of all faith beyond faith that we personally will live long enough to see the outcome of our daily activities. Most people do care what happens to their grandchildren and great-grandchildren, and in fact most now care about the preservation of Earth for our descendants. Is a line to be drawn after which it stops being important?

I think not. The essence of faith is the belief that there is pattern and purpose in the universe. If there is, if there's some point to existence, then the continued existence of our race does matter. And if there isn't, our own personal existence makes no difference either—we might as well all be dead. To be sure, there are a few who take the position that life is meaningless, but it is not a common one, and not one conducive to the accomplishment of anything of value.

The great majority of people, like Noren, have more underlying faith than they realize. Often in defiance of reason they feel, even if not consciously, that in the long run things will turn out well for the human race, whatever happens to them as individuals. Why else do some sacrifice their lives for the benefit of humankind? As to the ultimate fate of the individual, Noren confronts that question and is baffled by it, as most of us are; but in this too he senses that there must be truth beyond anyone's understanding. It is this awareness, the awareness that the universe has inherent pattern that we cannot comprehend, that constitutes faith. And this is what I hope my novels inspire readers to feel.

Update on the Critical Stage:
The Far Side of Evil's Relevance Today

This essay explains why I feel The Far Side of Evil *should not be called outdated, as it sometimes has been, and also presents my recent thoughts about the theory of the Critical Stage that underlies that novel.*

*

The central idea of my 1971 novel *The Far Side of Evil* is one that came to me 1956: my theory of the Critical Stage, the time between a planetary civilization's development of the means to destroy itself and a commitment to expand beyond the single planet where such destruction would wipe it out. The same level of technology that makes one possible also permits the other, and in my view they are mutually exclusive alternatives—a world will remain in the Critical Stage until one or the other happens. This supposition is as valid in my eyes as ever, although my conception of the Critical Stage has changed over time.

It's frustrating to me that many readers feel that the novel is outdated, and that it therefore seems irrelevant to today's world. I addressed this in the Afterword to the 2003 revised edition, but some people say the new version, too, is dated, so evidently I failed to revise the original text successfully. That's too bad, as I feel the book is even more relevant today than at the time of its initial publication.

The political situation in the story was never meant to parallel current events on Earth; it is comparable to our world as it was during the early fifties, not the seventies when the first edition appeared. After all, I wrote an initial draft of portions of it a year before the launch of Sputnik, an event that to my surprise and relief made it impossible for the planet portrayed to be our own. The story is not about politics, although its setting—the conflict between dictatorship and freedom—is universal and applies to all eras. As far as the story is concerned, that conflict is merely a plot device; so the fact that we no longer have two superpowers on the verge of nuclear war in no way dates it. As I said in the Afterword, some readers thought I used space fiction as a vehicle for political commentary when in fact it was the other way around: I used political melodrama to dramatize my ideas about the importance of traveling into space.

When the book was written, I assumed a world's Critical Stage is short. (Yes, I believe the theory applies to worlds other than ours, just as some scientists now believe that one goal of seeking interstellar radio contact is to find out how long an average planetary civilization lasts before self-destruction.) At the time of the Apollo moon landings, most people thought that nuclear war was likely to occur in the near future, but that if it didn't, we would continue to make rapid progress in space exploration. Since personally, I had believed since the early fifties that devoting a society's energy to space travel puts an end to the danger of a catastrophic nuclear war, I described the Critical Stage in those terms. And in fact, some evidence was provided by the space race with the Soviets, which absorbed money and effort that would otherwise have been spent on a more destructive competition.

As time passed, however, it became clear that my theory was a gross oversimplification. I tried to update it in the 2003 edition, pointing out that there are more dangers to a planetary civilization than nuclear war and that mere development of space travel capability, without a major commitment to establish settlements on other worlds, is not enough to eliminate them. But it's a novel, not a philosophic treatise, and it was being issued by the publisher's "children's book" department (although it's inappropriate for readers below high school age), so I wasn't able to elaborate enough to clarify the relevance to today's world.

Readers say to me that we have space travel yet are still in danger, and that's true. But we haven't made use of our space travel capability. Expansion into space prevents a civilization's destruction by two means: first, by constructively channeling the energy that would otherwise have gone into war, and second, by establishing footholds that can survive even if a species' home world is devastated—which in principle can happen through a natural event such as an asteroid strike, as well as through various kinds of human action. We have not taken steps toward either one; for nearly half a century, despite the dedicated effort of a small number of astronauts and space advocates, we have done no more than maintain a limited human presence in low orbit. Society as a whole has made no effort at all.

One reader told me he felt that we should spend no more on space travel until every child on Earth is well fed. That dismays me, as I believe the all-too-common idea that we should solve the problems on Earth before moving outward into space is a self-defeating policy. If we wait until we have eradicated poverty to colonize other worlds, neither will ever happen. Expansion into space is the solution—and in my opinion, the only solution—to Earth's problems. Abolishing hunger and pollution and war depends on the use of extraterrestrial resources. The fact that these problems still exist despite well-intentioned efforts to eliminate them is the result of our confinement to a single planet that we have outgrown, and they will inevitably continue to worsen until we make the effort to expand our civilization beyond it.

Most people assume either that we will someday learn to prevent war, or that human nature will eventually lead to interplanetary wars. In my opinion neither of those things will occur. War cannot be abolished by "learning" to prevent it. Negotiation is meaningless because no matter how many leaders negotiate in good faith, there will be fanatics who ignore treaties, and as long as these fanatics can attract enough followers to launch attacks, a strong defense against them is essential; failure to maintain it would lead to worldwide dictatorship. I believe that in time there will be an end to war, but this cannot happen until it becomes impossible for aggressors to recruit a significant number of supporters, a situation that can be brought about only by eliminating the factors that cause people to support them.

Human nature leads to war for two reasons that can't be merely wished away (although the majority attitude toward war certainly becomes more negative as the centuries pass). In the first place, humans crave challenge and excitement, which is the reason our species has been able to survive and thrive; so when a society is not fully occupied with a constructive challenge, it fulfills this need through a destructive one. In the second place, people fight over land and resources when these are scarce or seem likely to become scarce—again, this is an instinct indispensable to survival.

But once we break free of the confines of our native planet, there will be plenty of constructive challenge in the process of surviving elsewhere, and neither living space nor resources will ever be scarce again. The universe contains sufficient resources to last virtually forever—and making use of them needn't involve stealing them from extraterrestrial races; the discovery of numerous exoplanets indicates that there are more than enough uninhabited worlds to go around.

As long we are bound to a single world with shrinking resources, however, the situation can only get worse. There is nothing surprising in the rise of militant groups and terrorists; how could it be otherwise when it's obvious that Earth's resources can't last indefinitely and some societies either have less than others, or fear that what they do have will be taken away? When the frustration of the have-nots, and their lack of any way of constructively changing their situation, makes them easy prey for fanatics who know all too well how to satisfy their instinctive longing for an exciting challenge? This was always true, but in the past trouble wasn't widespread enough to threaten the existence of Earth's civilization as a whole. With modern technology, it becomes increasingly possible for a small minority to endanger the entire planet, even without the use of nuclear weapons (or with them; it no longer takes a superpower to launch a nuclear attack). Yet with that same level of technology we could extend civilization beyond the planet so that even if the worst should happen here, our species will not be wiped out.

This is what the Critical Stage is, and far from being an outdated concept, it becomes more and more pertinent year by year. I see this as a natural stage of evolution. We don't have wars because we are foolish or morally deficient (although there will always be individual evildoers), and we won't have them when our species is mature enough to take up the challenge of interplanetary expansion. I don't believe there will ever be interplanetary war, as many people think is inevitable in view of past history. Conditions will not be the same as in the past. In the terminology of anthropology, war in the past was adaptive for our species—it led step by step to the development of the technology needed to access the resources of a new ecological niche. It will not be adaptive once we are occupying that niche.

Nor will people's attitudes be the same. Centuries ago, war was considered glorious and men felt deprived of opportunity when no war was in progress. Even as recently as World War I, young Americans who joined up were afraid it would be over before they had a chance to get into the fight. Nobody in our culture feels that way today—it's generally agreed that war is a bad thing to be avoided whenever possible. Progress does occur over time. But it takes time—evolution is not a process that can be speeded up by decree, although it can, unfortunately, be stalled by apathy.

For many years I was increasingly worried, not so much by my awareness of more threats as by the fact that nothing was being done to speed up our progress in space and the general public cared less and less about it. I was afraid that our Critical Stage might be unnaturally prolonged. Then, a few years after the republication of *The Far Side of Evil,* it dawned on me that the public's decreasing interest in space is due not to apathy, but to fear—not conscious fear, but the stirring of an unconscious recognition that the universe is very much vaster, and more scary, than most people like to think. (See "Achieving Human Commitment to Space Colonization: Is Fear the Answer?" at my website,) At the time of Columbus, many thought venturesome ships would fall off the edge of the world, a prospect they viewed with great dismay; others (according to legend), knowing the world extended beyond their maps, marked the edges with the warning "Here Be Dragons." Figuratively speaking, most people of our time, having been shown that travel between worlds is no mere fantasy, may feel the same way about space exploration.

And so for a while I thought that the alternative fear of such disasters as biological warfare, environmental deterioration, and terrorism might be the spur needed to get the space program moving again—we wouldn't have gotten to moon without the fear that the Soviets would win the Cold War. But in 2012, while revising my nonfiction book *The Planet-Girded Suns* for republication, I suddenly saw the striking parallel between today's widespread underlying fear of what the universe may hold and the feeling that prevailed in the seventeenth century when the orderly Earth-centered conception of the cosmos was being replaced by realization that the universe has no center and the stars aren't fixed to a solid crystal sphere. The deep feeling of insecurity this new outlook engendered among the majority of educated people lasted for nearly a hundred years. Is there any reason to assume it will take less time for the public to get used to awareness that humankind is not isolated from whatever exists elsewhere? (See "Confronting the Universe in the Twenty-First Century," published as an Afterword to *The Planet-Girded Suns* and by *The Space Review*.)

And so I now think that a commitment to large-scale space efforts will not come soon, and that far from being a sign that something has gone wrong, this is a normal phase of evolution that should have been predictable. The Critical Stage simply isn't as brief as I once believed. That's an optimistic view, as it means we are still on track. But of course the danger of self-destruction remains, and will last until we do take major steps toward space colonization. There is a longer period of peril than I supposed, and thus greater odds that we won't survive it.

In the novel the Service is searching for "the key to the Critical Stage" that might enable them to save other worlds, and when readers asked why I didn't let them find it, I've replied it was because I didn't know the key myself. I now suspect that I do know, and that the only key is time. Thus there is indeed one unreasonable premise underlying the story—the assumption that the Service wasn't already aware of that, considering that it knew the histories of the many worlds in the Federation it represented. However, I naturally don't pretend to portray the very advanced interstellar civilization in my fiction realistically, so I trust that this newly-discovered plot hole is not too serious a flaw. Certainly their immediate concern is valid, since if no start were made toward developing space technology, a world's Critical Stage would end sooner or later in disaster.

Of course there are many individuals in our society who don't share the prevalent uneasiness about human contact with the universe and are enthusiastic about exploring. There will be small-scale activity in space, including bases on the moon and Mars, long before our Critical Stage is over. We are already beginning to make progress with commercial space ventures, which I have always believed are what are needed to bring about significant development of extraterrestrial resources. But I no longer believe we will see any major effort toward colonization before the end of the twenty-first century.

What will happen when our world's Critical Stage finally ends? By definition, we will have begun to spread into space, and more resources will be available to Earth. But I don't think our world will become the utopia many people hope for until much later; I suspect that for the foreseeable future such a society will be possible only in the colonies. I do believe war will be abandoned, yet there will still be troublemakers and police will be needed to deal with them. There will be hunger and poverty because Earth will be overcrowded for a long time to come, and it won't be possible to import sufficient resources until orbiting colonies—as distinguished from those on other planets—are well established (although implementation of space-based solar power could go a long way toward minimizing these problems). And there will be depression and apathy among the majority of citizens who cannot personally participate in the exploration and settling of the frontier. The novels of my *Captain of Estel* trilogy, which are set long after many worlds have been colonized, portray what I think is most likely; they paint a dismal picture of conditions on Earth but offer hope from an unexpected direction at the end of the story.

Today's space enthusiasts naturally resist the idea of there being a natural explanation for the slowness of our movement beyond Earth, one that time alone can overcome. My published essay about it wasn't warmly received. Perhaps I am able to believe it only because now that I'm past eighty I know colonies can't be established in my lifetime anyway, nor will I live to see the worsening conditions on Earth likely to prevail before they are. But if my theory about the delay is true, at least we have no present cause to doubt that, barring catastrophe, we will someday reach that pivotal point in our evolution.

A Response to Some Reactions to *Journey Between Worlds*

In 2015 I got the rights to Journey Between Worlds *back from the publisher of the 2007 paperback edition and issued ebook editions to replace theirs, which was overpriced. I hoped that at last it would reach the audience for which it was intended, but so far that hasn't happened; there just doesn't seem to be any way to make the readers most likely to enjoy it aware that it exists. Moreover, no one who has read it seems to have noticed the symbolism in the characterization of the story's heroine.*

*

My Young Adult novel *Journey Between Worlds*, a realistic story about the colonization of Mars, is about ordinary people living on that world. It's enjoyed more by young women who like romance than by avid sci-fi fans—while classed as "science fiction" because of its interplanetary setting, it's not about technology or exotic adventure. It's mainly a story about human aspirations and human love.

Journey Between Worlds was my first book, written before *Enchantress from the Stars* although published later. Admirers of *Enchantress* looking for something comparable tend to be disappointed in it, because it's not at all similar; but why should it be? A writer is not obligated to direct all his or her work to the same people. I aim to express my ideas in a variety of ways, not to meet the public's expectations. In theory, this should give me a wider audience. In practice, few of the readers for whom they were intended discover my lesser-known novels.

Journey has had two hardcover editions and a paperback issued by major publishers, yet despite good reviews it has never reached the audience it was meant for. The publishers marketed it exclusively as science fiction rather than romance, yet science fiction fans aren't likely to sympathize with a heroine who doesn't want to go to Mars. And they aren't the readers who need to be convinced that colonizing Mars is worthwhile. My aim was to appeal to romance readers who may or may not be in favor of space travel, and who wonder whether in a changing world, there's hope for their descendants to find happiness. And in fact, the 2006 hardcover edition was well received by the romance websites to which I personally arranged for it to be sent; some reviewers said they enjoyed it as adults as well as recommending it for teens, and even that it made them think.

Of course space enthusiasts interested in Martian colonies also like the book. The National Space Society review of the 2006 hardcover edition called it "A must-read for all future space pioneers who wish to persuade their friends to join them in making that future journey between the worlds of the known and the unknown." But it has remained largely invisible to the non-pioneers to whom I believe it would most appeal.

Comments at Amazon, Goodreads, and Facebook show that many readers who did find the book loved it and remember it as one of their favorites. Naturally there are critical comments, too, and where these concern the reader's personal lack of sympathy for the heroine, that's okay. If a character is portrayed as anything but perfect, not everybody will like her (and after all, if she were perfect, the book would be criticized as unrealistic). In this case, some found her particularly unlikable because her attitudes were different from their own and from those most prevalent in today's society.

But though I don't usually comment on negative reviews, there were several criticisms to which I feel compelled to reply. In the first place, many readers said that the heroine doesn't seem like a modern woman and that her outlook is sexist and obsolete. In the 1970 edition, this is certainly true. I wrote the book in the late sixties, and it did indeed reflect the views still common in that era—girls wanted mainly to get married and weren't very ambitious with regard to their careers. A lot of revision was done for the 2006 edition to eliminate dated phrasing and assumptions, as well as to update the description of the Martian landscape and some minor references to pre-computer technology that is now obsolete.

Yet despite the elimination of sexist attitudes, a number of reviewers complained that in the 2006 edition Melinda still doesn't seem like a modern woman, let alone like a woman of the future—and though this too is true, it's not a matter of when the book was written. It is an intentional portrayal of her individual personality. She is shy and dependent (although it's *not* true as some asserted that she lets her boyfriend control her; on the contrary, her rebellion against his attempt to tell her she can't go to Mars is the deciding factor in her decision to go). If she were typical of her generation she'd be eager for a trip to Mars! Moreover, there would be no plot conflict, no room for the growth and change that's the main point of the story. Apparently some of these readers never finished the book (a few stated that they didn't) because by the end of it, her views have changed. Her basic personality hasn't. She still wants marriage and a home—as most women do today, even when it's not their prime objective. She is still somewhat homesick for Earth, as the average person would be. But she's looking forward, not back, and her career plans have become more ambitious and more important to her.

If I were writing the book today I would not change Melinda's personality. It is, or is supposed to be, more significant than a mere plot device—it's basic to the book's theme. There is symbolism that my readers, even those who praised the book, seem to have missed. (I'm sometimes accused of spelling ideas out too much, but when I don't make them explicit, they don't get across. I guess I don't know how to be successfully subtle.) Melinda initially clings to what previous generations considered normal and natural, what she has always believed she wants, and sees no need to move beyond the confines of her past experience. So too do most people today cling to the idea that confinement to Earth is "natural" and see no necessity for the human race to go beyond the limits of the world where our ancestors evolved. I doubt if any such people grasped this parallel, or even that space supporters who didn't like Melinda did; but it is, in my opinion, quite exact. In both cases the underlying factor is a deep-seated longing for stability and fear of the unknown.

This bring me to the other criticism of the book that I want to comment on. One reviewer said, "The author has a definite bias in favor of space colonization." Well, if this is bias I plead guilty. But "bias" is not the right word. We don't normally call deeply held convictions "bias." We wouldn't say authors of anti-war novels are biased against war, or that advocates of banning industrial pollution are biased in favor of protecting the environment. Open supporters of a cause don't claim to be impartial, and no book other than a nonfiction survey of an issue should be expected to give weight to views unlike its author's. I make no secret of the fact that I believe space colonization is essential to the long-term survival of the human race. I recognize that not all readers share my opinion, and it's their right to disagree with it. But to imply that I ought to present both sides objectively in a novel is to ignore the difference between educational material and literature.

Today, in 2017 when there is more public interest in Mars than in the past and SpaceX CEO Elon Musk is actually planning to send people there, *Journey Between Worlds* is more relevant than ever. Do I believe Martian colonies will be anything like the one in the story? Of course not—not in this century and perhaps not in the next. Early colonies will be small and the residents will undergo great hardship; by the time larger ones are established, lifestyles even on Earth will be very different from what I have described. But I believe the underlying feelings of my characters are a valid portrayal of how colonists will view their society. Like pioneers of every past era, they will think of their lives as normal, however unnatural their goals may seem to others. They will be ordinary people, doing what humans have always done to build a better future for themselves and those who come after them.

Breaking Out from Earth's Shell

Long ago people literally believed that an invisible, transparent shell surrounded Earth to hold up the stars. After that theory was discarded, the shell remained in a figurative sense, for Earth was thought to be forever isolated from the rest of the universe. Some people are comforted even today by this illusion, although it's time for humankind to break free.

*

Since the dawn of history humans have been drawn to the idea of reaching the heavens. Most cultures' gods were presumed to dwell there. The stars have always symbolized mankind's highest aspirations. They were, however, viewed as inaccessible, except perhaps by souls after death.

According to the worldview universally accepted in Europe prior to the seventeenth century, the realm of the fixed stars (as distinguished from planets) was literally perfect and unchanging. The stars were thought to be embedded in an invisible sphere—a shell—that encircled Earth, for if they were not attached to something, would they not fall? Contrary to a common misconception, the theory that Earth is not the center of the solar system. was not resisted because of any sense of demotion from the place of prime importance, or even because of conflict with the Bible—actually, it would not have aroused much opposition if the authorities of the time had not realized, more than fifty years after its publication, that it opened the door to questioning the nature of the stars. Copernicus himself never doubted the accepted theory; to him, they were still firmly attached to a crystal sphere surrounding the sun.

However, near the end of the sixteenth century the philosopher Giordano Bruno, a strong advocate of the Copernican theory, suggested for the first time that the stars are not mere lights in the sky, but suns with planets of their own. Although that was not his only heresy, many scholars believe it was the primary reason why he was burned at the stake, and why, after his books were banned, supporters of his astronomical ideas called themselves Copernicans instead of mentioning his name.

Rearrangement of the solar system was a relatively minor issue compared to the upheaval in both science and religion caused by denial of the stars' perfection.

To people who had believed themselves safely enclosed within a perfect sphere, beyond which lay Heaven, the idea of a universe full of suns at random distance from Earth was extremely upsetting. As John Donne put it in these famous lines from his poem "An Anatomy of the World" (1611), it removed all coherence from their worldview.

And new philosophy calls all in doubt,
The element of fire is quite put out,
The sun is lost, and th'earth, and no man's wit
Can well direct him where to look for it.
And freely men confess that this world's spent,
When in the planets and the firmament
They seek so many new; they see that this
Is crumbled out again to his atomies.
'Tis all in pieces, all coherence gone,
All just supply, and all relation.

This was not a foolish or naive reaction. Human thought is dependent on a stable foundation on which to rely. Psychologically, people are cast adrift if their basic premises are questioned, and this is an adaptive trait since without anchors it would be impossible for a society to function. Change must come gradually, through the exceptional few who are able to discard the outlook of their contemporaries. Once they do, a new outlook spreads—but that takes time.

Over time, people became reconciled to the loss of an immutable order in the universe, and by the late seventeenth century those with enough education to care about astronomy envisioned countless suns, all surrounded by planets which, like those of our own solar system, were assumed to be inhabited. Some thought they were the homes of angels or the souls of the dead, but the belief that they were inhabited by mortals superior to ourselves soon predominated. The one thing everyone agreed about was that they were not without tenants. It had formerly been believed that the heavenly bodies existed for the benefit of mankind, so since distant suns and planets were of no apparent benefit to us, it was reasoned that they must have been made for other mortals. It was taken for granted that God would not have created a "useless" world. This was not questioned until the middle of the nineteenth century, when after hot debate the conviction that all extrasolar worlds have inhabitants still prevailed. Not until early in the twentieth was it abandoned.

Ironically, we now know that uninhabited planets are not useless—it may well become possible for us to colonize them. They may prove to be our salvation when the resources of Earth are gone. But of course in earlier centuries that possibility did not occur to anyone.

For more than three hundred years it was believed, by educated people at least, that solar systems similar to ours exist. This is known because they are referred to in the writings not just of astronomers but of many well-known people such as Benjamin Franklin as well as in the popular magazines of the day, sermons, and even in textbooks for children. Also, a great deal of poetry, some of it book-length, was written about spectacular suns and their planets. Imaginary voyages through space appear in even in poems by major poets such as Milton, Shelley, and Byron.

These were spiritual voyages, not journeys in spaceships. For example, at the time of Newton's death it was often suggested that he might see distant planets on his way to Heaven. Many people longed to visit the worlds so frequently talked about, and doing so in an afterlife was the only route they could imagine. As late as the 1870s the American poet Henry Abbey wrote:

> Death, that dread annulment which life shuns,
> Or fain would shun, becomes to life the way,
> The thoroughfare to greater worlds on high,
> The bridge from star to star. Seek how we may,
> There is no other road across the sky;
> And, looking up, I hear star-voices say:
> "You could not reach us if you did not die."

But the longing for a closer look at other worlds was not shared by everyone. Searches for information about it turn up only what was written by those who were interested in cosmic space. Most people who heard of distant solar systems may not have been interested, may even have been disturbed by the thought. The French philosopher Pascal famously wrote, "The eternal silence of these infinite spaces terrifies me," and he can hardly have seen alone in feeling that way. It was not an issue people needed to be concerned about. Earth was, after all, safely isolated from the larger universe, as far as they knew. No one supposed that there could ever be actual contact could between worlds; it was as if an invisible shell still enclosed our own..

It is likely that the poet William Wordsworth's feeling about space was—and still is—more typical than that of the space enthusiasts. He was knowledgeable about astronomy and enjoyed watching the stars with his sister and friends. But he is best known for his love of nature. When in his famous poem "Tintern Abbey" (1798) he wrote:

> Therefore am I still
> A lover of the meadows and the woods,
> And mountains; and of all that we behold
> From this green earth. . . .

was he considering the stars a part of nature? There is no indication in the poem that he was, yet it seems unlikely that he would have used the phrase "*from* this green earth" rather than the more common "*on* this green earth" if he had never looked outward, thinking of Earth as part of the larger natural universe. And in fact in another poem be wrote, "The stars are mansions built by Nature's hand."

Be that as it may, it was Earth alone that he cared about. In "Peter Bell," describing his return from a fantasy space journey, Wordsworth revealed an outlook that is shared by many today, more than two hundred years later.

Swift Mercury resounds with mirth,
Great Jove is full of stately bowers;
But these, and all that they contain,
What are they to that tiny grain,
That darling speck of ours!

Then back to Earth, the dear green Earth;
Whole ages if I here should roam,
The world for my remarks and me
Would not a whit the better be;
I've left my heart at home.

See! there she is, the matchless Earth!
There spreads the fam'd Pacific Ocean!
Old Andes thrusts yon craggy spear
Through the grey clouds—the Alps are here
Like waters in commotion. . . .

And see the town where I was born!
Around those happy fields we span
In boyish gambols—I was lost
Where I have been, but on this coast
I feel I am a man.

The last few lines say it all: for the vast majority of people, their very identity depends on their presence on Earth. To leave it in fantasy is one thing, but to be aware that people can really leave, really venture into unknown regions, puts a whole new face on facts that have been known for centuries. And when astronauts do leave, even they are often more deeply moved by the sight of "the dear green earth" from a distance than by the beauty of the stars. Though astronauts are exceptional individuals who enjoy space flight and long to explore the universe, those are not the emotions the public vicariously shares.

The farther we go in space, the further removed space travel is from theory, the more evident this will become. In Wordsworth's time and for two centuries longer, the idea that there might be peril in space didn't occur to people. Their knowledge of the universe was very abstract. Even the few brief mentions of traveling to the stars that appeared in the late nineteenth century did not suggest that it would be dangerous, and certainly there was no suspicion that the presumed inhabitants of other worlds might not be friendly. That notion was introduced by H. G. Wells' 1897 novel *The War of the Worlds,* which was viewed as pure fantasy until a 1938 radio dramatization was broadcast in the form of a news report, nearly causing a panic. Science fiction of the 1920s was read only by those especially interested in it. However, starting in the 1930s, the hit comic strip and radio adventures of Buck Rogers forever changed the public's perception of space. It became the scene of violent action and exciting new concepts, and the development of V-2 rockets in World War II led to a suspicion that there might possibly be some truth in them.

Thus the first UFO sightings, which occurred in 1947 and were immediately associated with extraterrestrials, were followed by countless alien invasion films in the 1950s. It is often said that these movies were actually about the Cold War, and no doubt their plots were influenced by it; but I believe that underneath, they reflected the public's new realization that space may hold terrors. These films featured ridiculously-portrayed aliens and some were intentionally humorous, which suggests that viewers wanted to think that the whole idea of danger from beyond Earth was silly. When the development of satellite technology began to show that space travel is not silly, their popularity waned. People turned their attention to the competition in space, which really was connected to the Cold War. After America won they could no longer be distracted by it, and the worries suppressed so long began to surface along with growing anxiety about our own ventures into the unknown.

At the time of the Challenger disaster I was astonished by the widespread public feeling that it meant space travel shouldn't be undertaken, and especially that the civilian teacher Christa McAuliffe shouldn't have been "sent" into space, as if she hadn't been chosen out of thousands of applicants who vied for the chance to go. The supposition, sometimes even explicitly stated, that she hadn't known it was dangerous was an insult both to her courage and to her intelligence. Who could possibly be unaware that riding in a spacecraft propelled by rocket engines and boosters providing 7.8 million pounds of thrust at liftoff involves risk?

Perhaps previous space flights had been viewed with a sense of detachment, as if they were science fiction. But I now think there was more to it than that. It would simply not have been rational for anyone ever to think space travel isn't dangerous; the evidence that it's unsafe could hardly have come as a surprise. And even if it did, dangers involving far greater numbers of people, such as those of early aviation, had been accepted by the public without question. No one said that pilots shouldn't be allowed to take off in primitive planes, although the crash rate was extremely high. Planes, however, did not get very far from the ground. There was no possibility that improved ones would leave the planet and enter unknown regions beyond. I suspect that the realization that space travel is *real* came not with the tragedy of *Challenger*, but with the Apollo moon flights, and that *Challenger* brought to the surface unconscious feelings that had been building up for a long time. Underneath, people were troubled not by the danger to the astronauts but by the potential perils of contact with the wider universe. The Challenger disaster was merely the trigger for expression of the public's growing uneasiness about spacefaring.

As in the seventeenth century, for people to shrink from the necessity of revising their perception of our environment (our total environment, not the mere biosphere) is normal and adaptive for our species. If everyone's orientation shifted suddenly, society would disintegrate. Civilization depends on the ability to make plans knowing that for the short term, tomorrow will be like today. Thus changes in outlook come slowly, first in a few visionaries, later in one generation after another as minds open to new awareness. Mass media, a recent phenomenon, will speed up the process but cannot make it happen overnight. If anything, real-time mass media events such as the moon landings produce more shock than lasting transformation.

So it can't be expected that the public will be quick to

support future space activity. We should not be surprised if more interest is shown in science fiction movies than in real flights. Inwardly drawn to the thought of venturing outward but unable to break away from the safe and familiar, people tell themselves that the fiction they enjoy is just entertainment, not be taken seriously. And science fiction is sometimes criticized for promoting unrealistic dreams.

Will we always be bound to Earth, then? Of course not. Evolution is slow, but it can't be halted. Humans have been seeking new lands to settle since for millennia, first new villages, and eventually new continents. The negative expression of this instinct has been the urge to expand a group's territory through war, which hopefully most us have outgrown since the time when young men dreamed of glorious conquest. On the positive side, there has always been a desire of ordinary people to , even at the cost of hardship. For some time it has been evident that Earth has, or soon will, run out of vacant land. How could there not be an impulse to go beyond, quite apart from the plain fact that a species that fails to move beyond the niche in which it has evolved must be periodically decimated or else fall victim to extinction? In the long run, how could humankind fail to follow that impulse?

The dream of extending our species' range beyond the world on which we evolved is hardly something so trivial as entertainment, however much entertainment may be derived from it or how gradually it is absorbed. It is an often-unconscious expression of the deep-seated instinct present in all species to expand their ecological niche, an adaptive response to the ever-present threat of extinction. It has become trite and unfashionable to compare movement beyond Earth to the movement of life from the sea onto land, as was done during the Apollo era, but that comparison is still valid.

These are not new ideas—space advocates have been expressing them for years. Why then have so many lost sight of them and become discouraged? I think it is because in our era people are so used to rapid change, and to instant gratification of their wishes, that they have lost all sense of the evolutionary timescale. A dream is not unrealistic merely because it is not achieved within one's own lifetime or even that of one's grandchildren. Enthusiasm for one ambitious space dream after another has died out when its supporters came up against the fact that they wouldn't live to see it fulfilled—a reaction that strikes me as all too close to "sour grapes." As has often been pointed out, settling space is not as simple as picking up stakes and moving one's family westward. It requires a very long lead time. During that time, the clock would stop if there were no far-sighted dreamers willing to pay the price of personal disappointment in order to keep it going. The more followers they can attract by offering entertainment, the better; but to suppose that their motivation has no deeper roots is to ignore the essence of what enables our species, or any species, to thrive.

Sooner or later, like an eaglet destined for flight, humankind will break through the invisible shell in which our planet has been confined. It is happening now with the advent of commercial space flight, and the minority with far sight will carry us forward despite reluctance on the part of the majority. In times to come men and women will travel far from this green earth. And then, with our ancestral home at last open to the universe, we will discover our place among the countless worlds of the stars.

The Role of Psi in Human Affairs

This essay presents some of my ideas about how mythologies, among other things, spread through a society—ideas too unorthodox for me to go into in material for the course I taught on Space Age mythology. It also explains some of the ideas underlying my novels.

*

I was once told in an online forum that if independently verifiable results of scientific psi experiments were published in a respectable peer-reviewed journal, "even a smallest such result would be a revolutionary change in the understanding of the world." If only this were true. In fact many peer-reviewed studies have been published, and if even the smallest positive result could change most people's understanding, there would have been a major revolution in science—and I believe in our view of history—long ago. However, there are too many people who don't want to look at the results. They don't want to change their understanding of the world. Scientists in particular are aware that if psi exists, the universe doesn't work the way they think it does and they have built their careers on false premises. So they don't read the relevant journals or the books that cite them, and thus preserve the comfortable (to them) illusion that only mystics and charlatans believe in phenomena that can't be explained by materialistic science.

There is no doubt whatsoever that so-called paranormal (psi) abilities such as ESP do exist. This has been proven repeatedly by responsible scientific investigation and by use in once-secret military projects, quite apart from the extensive evidence from human history. But the issue is generally confused by all the other things, often silly things, associated in the popular mind with the terms "paranormal," "psychic," and "parapsychology," and by the use of the word "supernatural" in connection with psi phenomena. Those that are real aren't in any sense supernatural; they are simply natural human (and in some cases animal) capabilities that we don't yet understand.

Properly speaking, parapsychology is the scientific, usually academic, study of psi: that is, extrasensory perception (ESP)—telepathy, clairvoyance (now often called remote viewing), and precognition—and psychokinesis (PK). It generally does *not* involve investigation of ghosts, although some scientists do include the question of survival after death, which I personally consider a separate issue, within its purview.

Skeptics are fond of pointing out that many people who claim to be psychics are frauds. Of course they are; but there is, after all, plenty of fraud in other areas. This does not mean that no one has real psi abilities. We don't stop buying cars because some used car salesmen are crooks, nor do we distrust stockbrokers in general because some of them sell phony stocks. Genuine psychics with conscious control of their powers are rare, to be sure; most psi occurs unconsciously. In my opinion unconscious psi has been, and still is, vastly more prevalent than anyone imagines and has had a major impact on human history. Why isn't this recognized? Why do so many otherwise open-minded people, scientists in particular, vehemently reject evidence that in any other context would be indisputable? As Kira says in my novel *Stewards of the Flame*, it's not because they think it's nonsense, as they sincerely believe, but because underneath they are afraid that it's *not* nonsense.

"Over and over, scientific evidence was presented for the existence of telepathy and other psi powers," she tells Jesse. "By most people, especially other scientists, it was ignored, blocked out of consciousness—often angrily denied. Because if it exists, then everything familiar to you about your mind, your world, stands open to challenge. There's nothing firm left to cling to. And if you have paranormal powers, who knows what you might do to disrupt the world, unwittingly, perhaps even unwillingly?"

Eminent psi researcher Dean Radin says in the passage from his book *Supernormal*, which I used as an epigraph to *Herald of the Flame*: "Real psychic effects lurking in the dark boundaries between mind and matter are so frightening and disorienting that defense mechanisms immediately snap into place to protect our psyches from these disturbing thoughts. We become blind to personal psychic episodes and to the supportive scientific evidence, we conveniently forget mind-shattering synchronicities, and if the intensity of the mysterium tremendum becomes too hot, we angrily deny any interest in the topic while backing away and vigorously making the sign of the cross."

This reaction, though prevalent, is not widely recognized. And it is by no means the only obstacle to acceptance of psi. The major acknowledged factor is that there is as yet no theory to explain how it works. Psi violates all known laws of physics, so even people who believe in it are unable to see how it *can* work, though some researchers are now attempting to relate it to quantum physics. Personally, I think the basic premise that all reality can be reduced to physics, at the quantum level if not otherwise, is open to question; the fact that this is true of all phenomena so far understood tells nothing about what may be discovered in the future. Be that as it may, science has never accepted any area as worthy of study without some sort of theoretical foundation, and it is not likely to do so in the case of psi—not when scientists have built their knowledge of the universe on a materialistic concept of existence. When someone does come up with a viable theory, he or she will be classed by historians with Newton, Darwin, and Einstein; but meanwhile the evidence is being swept under the rug.

Another main factor in psi's rejection by scientists is that the use of paranormal abilities is not reliable. The results of lab tests are predicable only in the statistical sense; even the best psychics miss some of the time. This in itself would not be a serious objection, as many other scientific studies are based on statistics. But lab experiments cannot reveal more about psi than the fact that such abilities exist, since the controlled use of them requires more significant goals than pertain under lab conditions. It demands tasks of importance either to society or to the individual, and emotion of some kind must be involved. And even then, the information it reveals, while often true, is also often false. Remote viewers, for example, can see actual places or happenings, but some of what they see is not real—and they cannot tell the difference. Independent verification, which is generally hard to get on short notice, is always required. This was one reason (but not the only one) why the military "psychic spying" program of the 1980s was discontinued.

Still another cause of opposition to psi is its long association with the occult, New Age ideas, and fringe beliefs. Throughout history psi phenomena have been dealt with by practitioners in these areas, as well as by many religions, especially Eastern religions—all of which have well-established traditional metaphors for explaining and controlling these phenomena. Outsiders find such metaphors, such as mysterious forms of healing energy or an invisible "third eye" that can perceive things not detectable through the senses, extremely off-putting, yet psychics who have learned to function through the use of them insist that they exist. Very few people understand the nature of metaphor. Both believers and nonbelievers generally assume that either a thing is literally true or it is false. To me, belief in metaphorical descriptions of psi abilities is exactly the same as Georyn's belief in a dragon in *Enchantress from the Stars*—it is simply a way of perceiving truth that fits a person's background. But scientists and others who consider themselves rational dismiss such beliefs as wacky, and tend to throw out the baby with the bathwater.

This, too, was among the many reasons for the abandonment of the military remote viewing program. Several people associated with that program, along with a few other prominent remote viewers, began looking for—and seeing—such things as alien bases on the moon and other planets, or underground on Earth, or insisting that extraterrestrials are trying to contact us. While this is not a traditional metaphor, it is a very common one today, among UFO cultists and "abductees" among others. It's important to realize that many of these people are intelligent men and women who hold responsible positions, and are in no way mentally ill. They really did "see" alien bases. Psi works in this way; sometimes what's seen is real and other times it is not, and the brain cannot distinguish real visions from illusions.

The interesting question is, why do people perceive similar unreal scenes? It's usually said that in the past they believed in religious figures such as angels, and had seen pictures of angels, so that's what they psychically saw; while now they're familiar with the concept of aliens. That's true, but I think there is a good deal more to it than that. In the first place the details of these visions, particularly the visions of those who think they've been abducted by UFOs (but not necessarily those of experienced remote viewers), are more alike than can be accounted for by any pictures or descriptions the people involved have been exposed to. In the second place, I believe, though it's rarely if ever mentioned by analysts, that what's seen is in fact metaphorical—it means something more than it seems to on the surface.

That is why some people accustomed to remote viewing have pursued a search for aliens despite the apparent foolishness of the effort. Unlike the average person they cannot shut off the ability to see, and they are drawn to the metaphor now dominant in the collective unconscious. It seems especially real to them because many other minds are unconsciously turned in the same direction, toward a prevalent symbol of emotions that are now widespread. In my essay "The Significance of Belief in UFOs" I have suggested what these emotions may be. We live in an era of upheaval in human conception of the universe, now that the reality of space travel has made us aware that Earth is not isolated as has always been supposed—that not only may we someday enter unknown regions, but are already vulnerable to visitation from beyond. As I have explained in my essay "Confronting the Universe in the Twenty-First Century"(in *From This Green Earth*), this is comparable to the uneasiness prevalent in the seventeenth century when people were absorbing the idea that Earth is not the center of the universe and the stars are not mere lights fixed to a crystal sphere. In the general public this shift in orientation produces deep unconscious anxiety. On the other hand, a minority is elated by it, and some (including some scientists and well as fringe cults) hope for aid from advanced extraterrestrial civilizations. Either way, the belief that aliens are present is a powerful symbol of these emotions, whether or not it is consciously recognized as such. So that belief is absorbed by the collective unconscious and is the source of metaphorical visions seen by those whose who are psi-receptive and whose personal unconscious feelings are particularly strong.

*

This is an example of how ideas have spread from person to

person throughout human history. Psychiatrist Carl Jung, who introduced the concept of the collective unconscious, believed that its content is inborn in all humans, and since no one knows how it is transmitted, science generally holds that there is no such thing. Now, some people are saying that the collective unconscious consists of material spread throughout the population via unconscious telepathy, something that has always seemed to me self-evident. How else could thoughts spread, when there are countless cases that sensory transmission cannot account for? To Jung, the concepts held in common were limited to archetypes, but much more specific ideas are extensively shared; ideas have even been passed from one culture to another without any satisfactory anthropological explanation.

Almost all telepathy occurs on an unconscious level, which is why many people think that because they themselves aren't psychic, and aren't acquainted with anyone who is, ESP is rare if it exists at all. But parapsychologists believe that the capability is latent in everyone, and in my opinion it is active in everyone even though few are aware of it. To be consciously psychic is indeed rare, and the talent for it appears to vary widely, just as does musical or mathematical talent. Genetics probably plays a part in this. There are, however, many individuals who have occasional spontaneous psychic experiences, and many who do not but can nevertheless learn remote viewing. How much of the learning—and for that matter, the genetic aptitude—involves a special paranormal faculty and how much is simply a matter of ability to gain access to the content of one's unconscious mind? How much spontaneous psi is the result not of unprecedented input, but of momentary breakthroughs of emotionally-significant aspects of what the unconscious mind receives all the time?

It is well known that all so-called "right-brain" abilities—artistic creativity, spirituality, and intuition—arise from the unconscious mind, as do the roots of emotion. And conscious psi functioning, including psychic healing, requires the same kind of relaxation, the same letting go of "left-brain" logical thinking, as do these other activities; it is systematically practiced in preparation for all of them as well as in many religious traditions. Some people are better at this than others—at one extreme it comes naturally, while at the opposite one it's virtually impossible; but everyone has at least a minor degree of potential.

Thus it is quite possible that some of everyone's ideas and feelings are continually, though unconsciously, made available to everyone else. There are speculators who view this as a group mind or hive mind, implying or even stating that there is only one consciousness of which individual minds are mere illusory parts. I disagree strongly with that theory; among other flaws, it leaves no room for privacy. The fear that others might read their minds is a major factor in skeptics' reluctance to accept the reality of telepathy, and I think it is an unwarranted fear. Telepathy is not "mind reading" and individuals don't have unlimited access to each other's minds, either consciously or unconsciously. It's not necessary to postulate a group mind to explain the collective unconscious, for psi is known to be instantaneous and therefore no time is needed for transmission of ideas and images.

This is not to say that the experience, common in mysticism and altered states, of feeling one with universe is not real. Certainly it is possible to merge one's consciousness with others, and with something beyond human consciousness, so the awareness of individuality is temporarily suspended. But to me this is a matter of connection, not absorption. However closely connected minds may be, they retain the ability to function separately as long as they are alive; after that, no one knows, but I suspect personality exists in whatever form of afterlife there may be—something skeptics are more prone to believe in the case of loved ones than in their own case. Whether or not that's true, psi in ordinary life involves communication either between one person to another, or more routinely to people in general.

It's my belief that individuals communicate what they want, underneath, to share—both things they feel strongly about and background details of their surroundings and situation. Who has not longed to make others see what he or she is truly feeling, either to be understood as a person or to convince them of the rightness or wrongness of something? There is more than enough such longing among the population to fill the collective unconscious without resorting to a denial of individuality that precludes, among other things, personal bonds of love as distinguished from general love of all humankind.

And so I believe that continuously, since ancient times, the feelings and convictions of the public about what matters in their lives have poured into a pool from which everyone can draw. By this means thoughts and visual images that are most common in a society are adopted by its members and become still more common. This leads, of course, to the spread of evil as well as good. Phenomena such as mass hysteria and mass racial hatred are universally acknowledged although their telepathic origin is not. And trivial enthusiasms such as fads are spread along with serious ones. But on the whole, the fears, hopes, and dreams of a culture are shared to an extent that has never been adequately explained under the assumption that sensory transmission is necessary.

In my opinion the ideas and imagery underlying mythology, as well as all the general knowledge taken for granted in a society including its concepts of right and wrong, is transmitted and absorbed in this way. The process begins in infancy, perhaps even earlier since I believe there is a telepathic bond between a mother and her unborn child. If this is true it sheds light on questions that are never raised because the answers are assumed to be obvious. Yet the sense of obviousness is not an inborn quality, nor is it usually the result of specific teaching. To take one example, why are young children who have never seen or touched live animals "naturally" attracted to mere pictures of them? What concept can a 3-year-old have of a cat or dog, let alone a lamb, before being exposed at least to a realistic movie? It can only come from the collective unconscious, formed by the love of animals felt by past generations.

Occasionally specific atypical feelings are drawn from this source. My mother told me that when I was small I was terrified by a picture of a goat. Neither I nor anyone I'd met had ever seen goats or heard anything bad about them, and I had no negative reaction to pictures of sheep, yet my fright was persistent. There are people who would say I must have been injured by a goat in a former life, but it's more consistent with what we know about psi to postulate that such seemingly-inexplicable impressions are derived from a common store of strong feelings once experienced, and telepathically projected, by individuals who either lived in the past or are still alive. And in fact, in some occult mythologies goats are associated with the devil, which would surely have led to such feelings.

Telepathy also accounts, I think, for the influence of great leaders, those who have inspired larger numbers of followers than can be accounted for by their outward difference from their peers. I believe they were telepathically gifted, whether or not they were aware of it, and projected their messages beneath the surface as well as through their words. This too has served both good and evil ends. At one extreme were the founders of major religions, who exemplify my belief that strong psi power is an advanced capability that in the future will be accompanied by moral advancement. To a lesser extent it applies to charismatic military or political leaders as well as entertainers such as rock stars who don't appear to have any real talent beyond the ability to attract fans. And on the negative end of the scale lie rabble-rousers and rulers such as Attila the Hun and Adolf Hitler, whose acquisition of a mass following despite his lack of any prepossessing physical or mental characteristics has always been a mystery. I am not saying that any of these people used telepathy intentionally or were conscious of their talent for it; only looking back in the light of what is now known about psi can its significance to history be considered.

*

Will we ever be able to use psi capabilities consciously to a greater extent than is now done by a few exceptionally gifted psychics and trained remote viewers? I believe so, but this lies far in the future; it will be a major step in evolution, not, as some people think, a return to powers lost in the past. My original idea for my Flame novels was to explore how, and why, a civilization might begin to move from the present level of ours to that of Elana's people in *Enchantress from the Stars*, who had very advanced psi powers. It would be a long process, taking place over many generations, but it would have to start somewhere. And it would not be adaptive in the evolutionary sense for it to become widespread before a civilization has developed the technology necessary for expansion to many worlds. To turn to "inner space" before spreading into outer space, as some people advocate, would be self-defeating, since colonization of space is essential to a species' long-term survival and to draw society's focus away from logical "left brain" thinking would distract us from that.

Actually, lack of adaptiveness is the reason for our not having more psi capability at present. If humans already possess the genetic capacity for paranormal development—as is shown by the cases where it has been proven to occur—this capacity must have evolved biologically at some time in the past. There is every reason to think that it did so before our separation from our hominid ancestors, and in fact before the appearance of hominids, since ESP in animals is known to be common. (Though it's beyond the scope of this short essay, not only have strong telepathic bonds between people and their pets been frequently demonstrated, but studies of wild animal behavior suggest that some aspects of it are the result of psi.) It is known, too, that psi faculties have been, and still are, used more in low-technology cultures—and presumably were in prehistoric ones—than in modern society. But the development of technology and the "left brain" rational mode of thinking it requires enabled our species to survive and thrive, and was therefore adaptive throughout most of human history. Thus "right-brain" modes of thought were suppressed and our primitive psi capabilities were largely forgotten. As early parapsychologist J. B. Rhine wrote, "psi is an elementary mode of reaction of the organism, one that probably represents the beginning of orientation in the initial adaptation to the environment," since it has been found to be inhibited by "the more recently acquired intellectual powers."

Why, then, do I view future development of human psi faculties as an evolutionary advance? Partly because I'm envisioning far greater paranormal powers than those of animals or our primitive ancestors, which I believe will be consciously employed. Conversational telepathy is one example, but psycho-kinesis too will become usable, as well as precognition, which has been shown to exist not only through lab experiments but through the remote viewing that has already been done. Whether people will ever be able to handle fire as they do in the Flame novels, I don't know; I used that primarily as a symbol of abilities greater than we now imagine—still, there have been reports of its occurrence in the past, and firewalking is common even today among people without other paranormal abilities.

Evolution, however, is the result of adaptation, and what adaptive advantage would there be such capabilities? For one, there is surely survival value in becoming consciously aware of what our fellow-humans are thinking and feeling, which will greatly reduce the misunderstanding and lack of empathy that leads not only to unnecessary suffering, but to conflict. Conflict is a greater danger now than it was when it was merely a matter of fighting among small tribes. We may find psi indispensable in preventing it.

Another possible advantage—as was suggested by Robert Heinlein in his classic novel *Time for the Stars,* telepathy may be useful for communication with starships traveling much faster than electronic signals. Psi is independent of time and space. Possibly it will prove to be the only means through which interstellar colonies can communicate with each other and with Earth.

Or perhaps, if artificial intelligence develops to the point where it exceeds human intelligence as many researchers think it may, psi capability will be what distinguishes us from robots and gives us a competitive advantage over them. Since it is not in itself a function of the brain—though the brain may act as a filter to narrow what a person draws from the immense store of data in the collective unconscious—it is a power no artificial intelligence will ever have.

A more immediate value of psi is its role in healing—and, as described in my Flame novels, in preventing illness. The kinds of illness with which medical science can deal are limited. Contrary to its current premises, bodies are not machines that can always, at least in principle, be repaired. Much that goes wrong with them is caused by disturbance in the unconscious mind, so that when one illness is eliminated another arises to take its place, and some conditions are incurable. The characters in the novels gain volitional control of their unconscious minds, which enables them to stay healthy. The extent to which this, and/or psychic healing, involves psychokinesis is not specified; we simply don't know how thought affects the brain's production of neurotransmitters, and I chose not to include an imaginary explanation. But I postulate that unconscious telepathic communication with the instructor is needed to learn volitional control, and that in some, this leads to conscious use of ESP, so I suspect that PK ability is also awakened.

Past evolution of the human mind has shown a steady trend toward increased volition in one area after another—that is, action through the exercise of deliberate, conscious choice. As I said in my book-review essay "The Evolutionary Significance of the Metanormal" (available at my website), "What animals do, they do instinctively. The assertions of strict behaviorists notwithstanding, humans learn to choose; and the ability to choose is what has enabled psychosocial evolution—from the earliest divisions of labor to the most complex achievements of modern technology—to take place. If we had not gained control first over our day-to-day actions and then over our rational minds, we would still be living in the manner of our hominid ancestors. That we have not learned to control our emotional responses at the same pace is a truism, but against those who assert that this indicates some sort of pathology, I have always maintained that it is a natural consequence of our evolutionary stage. Emotional faculties, like other unconscious functions, are not as easy to control as rational ones; a different form of volition is involved, probably different even at the neurological level. Everything we have discovered about metanormal capabilities indicates that they too involve this other form of volition and that conventional "willpower" . . . is counterproductive in the actual attainment of non-rational skills. This is as well known to athletes as to mystics. . . . We may not unreasonably suppose that the survival value of such skills lies less in their specific utility than in the new form of volition underlying them, and that this may in the future have as much effect on the extension of human powers as our control over rationality has had."

Personally, I believe that a major adaptive advantage in the controlled use of psi is that it's necessary to enable us to meet advanced extraterrestrials on terms of equality, without language-based misunderstanding, and that they will not reveal themselves to us until we have gained it. This is a premise of all five of my novels dealing with extraterrestrial civilizations. But to offer it as an explanation for adaptiveness begs the question, since it must for some reason have been adaptive for the extraterrestrial species that preceded us.

We really don't really know how psi will prove adaptive, any more than primitive humans knew where their newly-discovered reasoning powers would lead. Adaptiveness in the non-genetic sense involves long periods of time and can be perceived only with hindsight. But I think it's safe to say that any increase in the powers of human beings is in itself good, and that it may prove crucial in preparing us for whatever we may encounter as we explore aspects of the universe presently beyond our comprehension.

The Roots of Disbelief in Human Mind Powers

Originally, I planned to write two essays—one about scientists' misguided attempt to create conscious artificial intelligence, and the other about why most deny the overwhelming evidence for the existence of psi powers. But then I realized that these are really the same issue: the unwillingness to admit that human minds, and even animal minds, are more than brains.

*

Both scientists' conviction that AI can be made conscious and their denial of the evidence for psi arise from their unshakable belief in materialism, defined by the dictionary as "a theory that physical matter is the only or fundamental reality and that all being and processes and phenomena can be explained as manifestations or results of matter." And the primary root of refusal to question this belief is fear.

Why would scientists fear to question anything? Isn't an inquiring mind one of the main prerequisites for becoming a scientist? After all, discoveries in science are made by questioning past assumptions. Copernicus questioned the belief that the sun revolves around Earth. Darwin questioned the belief that species have separate origins, and Einstein questioned the belief that Newton's laws cover all circumstances. These, however, were exceptional men. Although lesser discoveries are also made by questioning, a lot depends on how widely applicable a belief is and how firmly established it has become. And the belief in materialism is so deeply entrenched in science that to question it shakes the very foundation of a scientist's professional life. That the universe may not work the way one believes it does is a frightening idea to entertain; few people are willing to let it into their thoughts.

Because of this, and because mavericks in any group are generally looked down upon, a scientist who does question a widely-accepted theory is likely to suffer the loss of professional reputation. This is another risk few are willing to take. It can mean ridicule, the disregard of otherwise-important work, and even being fired from a job. Professors, for example, are expected to teach the theories prevalent in their fields. Although in principle they have academic freedom, that doesn't mean much if they don't have tenure, and even if they do have it some excuse can often be found for dismissing a nonconformist. Students must go along with what they are taught if they except to get their degrees. Employees of research institutions must retain the respect of their colleagues. Thus the cards are stacked against anyone who attempts to challenge conventional ideas. As Voltaire wrote, "It is dangerous to be right in matters on which the established authorities are wrong."

But scientists' faith in materialism goes much deeper than this. They have a compelling desire to understand the world—that is why they chose to become scientists. Yet if they believe in science to the exclusion of other faiths, as many do, there is no current alternative to the materialist worldview. To reject it means to be cast adrift. We do not know anything about reality other than its material aspects unless we have had personal experience with another side of it or accept the word of those who have; it is beyond our understanding at this stage of our evolution. This does not mean there is anything supernatural about it. After all, two hundred years ago quantum physics was beyond comprehension; the human mind evolves. In time we will understand more. But in the meantime, to be confronted with something impossible to know is unendurable for many people, even terrifying, if they are left with no firm ground to stand on.

This is true not just of scientists, but of everyone. In our society science is widely viewed as the final arbiter of truth, so the majority of people share its prevailing idea of the nature of reality. In the past, that wasn't the case; most—even many scientists—considered religion the authority, as some still do today. Then gradually, people turned away from religion, at least with respect to its support of the concept of nonphysical reality. It is now common to consider religion authoritative on moral issues alone. Some sects do reject science, taking an equally dogmatic position in opposition to it. On the other hand, some individuals find meaning in alternative forms of spirituality that offer closer contact with nonmaterial reality than do most traditional religions. But comparatively few people are comfortable without any source of intellectual authority at all. Psychologists call the ability to accept not feeling sure about what's true "tolerance of uncertainty." Though it is unhealthy to lack it in dealing with daily affairs, uncertainty in regard to the basic attributes of the universe is another matter, and being able tolerate it is the exception rather than the rule.

Before the twentieth century, materialism was not the only philosophy accepted by scientists. While some did view bodies as machines, others—known as vitalists—held that living organisms are fundamentally different from inanimate things because they contain some sort of "vital force" that cannot be reduced to physicochemical processes. When little was known of biology this seemed rather obvious; how could mere physical components produce thought and feeling? The nature of the vital force was a mystery, but so much else about living things was mysterious that ignorance of this aspect of them was considered problematic only by philosophers,. But once biology was further advanced, it appeared that physical factors could indeed explain everything about organisms, and there was no longer any need to tolerate the uncertainty inherent in vitalism. It was vehemently rejected by science and any suggestion that there might have been some truth in the concept was scorned, as it still is today.

Yet In fact, physical factors cannot explain everything about organisms, although many scientists claim they can; and neither vitalism nor similar concepts, such as the Chinese idea of "chi" or the New Age theory of "energy flow," can explain it either. Now, however, the question of what makes living things conscious is receiving more and more attention. and scientists who believe that they are well on the way to creating robots with human (or superior) intelligence frankly acknowledge this as the elephant in the room. Most are convinced that sooner or later they will figure it out without recourse to inexplicable forces. No wonder they are desperately trying to produce such a robot—subconsciously they must know that if they can't, that may mean that there is a fatal flaw in the materialist premises on which modern science is based.

*

What is this nonphysical aspect of mind that science ignores and a great many humans believe in, yet cannot understand? We don't have a word for it. Some call it soul, but that term is too closely associated with religion, and with belief in an afterlife, to be meaningful to people who aren't religious. In any case mind distinct from brain is a broader concept than is generally associated with soul, although the dictionary definition, "the immaterial essence, animating principle or actuating cause of an individual life" is not far off. The concept of mind encompasses capabilities as well as essence—intuition and creativity through which ideas not based on reasoning are formed, plus the capacity for spontaneous emotions such as love.

Mere "intelligence" can provide none of these capabilities, since it functions by proceeding logically from premises even when the logic is flawed. Robots, however far they advance from what was originally programmed into them, are limited by the process of logic because that is how machine intelligence works. They might someday express emotions derived from the logical implications of circumstances, such as fear, pleasure, or even grief. But love is not founded on logic; it simply exists. And it is closely tied to the nonmaterial connection between all beings believed by many to be a fundamental aspect of the universe.

Although we cannot define or explain its nature, there is plenty of evidence that the mind is more than just the brain. Apart from its obvious attributes mentioned above, it possesses psi capabilities such as ESP that have no physical basis. Proof of the existence of psi phenomena, from laboratory studies and military use as well as from centuries of human experience, is extensive and indisputable; I have discussed this in my essay "The Role of Psi in Human Affairs." Needless to say, the machine theory of human nature is incompatible with these phenomena. That's the fundamental reason why the reality of psi is so vehemently denied.

If adherents of materialism are disturbed by any suspicion of an inherent difference between human minds and robot minds, they are absolutely aghast at the concept of psi. A noted reviewer of a well-designed research study on ESP that had been submitted to the *Journal of Personality and Social Psychology* told colleagues, "Reading it made me physically unwell." Hypothetical distinctions between human and artificial intelligence can be easily set aside as long as AGI doesn't exist, but psi phenomena can be ignored only by subconsciously suppressing any awareness of their existence that circumstances stir up.

Most scientists make every effort to avoid such circumstances. They don't read papers or books by the responsible researchers in the field. They dismiss the huge amount of nonsense posted on the Internet by New Age enthusiasts, crackpots, and fraudulent psychics, reassured by the thought that anything connected with psi abilities must fall in the same category. Thus unless confronted with a case of personal experience, they can tell themselves that it is nothing but pseudoscience and not have to face the possibility that both their careers and their personal worldviews have been built on sand.

That's not the only reason skeptics find the thought of psi troubling. The average person in our society, especially a scientifically-oriented person, has a deep-seated, innate fear not merely of acknowledging psi powers but of possessing them. Questions such as *What if someone could read my thoughts? What might I do with such powers without meaning to?* lie below the surface even if not consciously asked. This is especially true of people who sense that they have buried psi ability. It is latent in everyone, but obviously some people have greater extrasensory receptivity than others. We do not know whether there is a genetic component in this—it may be merely the degree of access to one's unconscious mind rather than a special talent. In any case a person who is psi-gifted is no more abnormal than a gifted musician; but some feel it would be more of a curse than a gift.

Even people utterly convinced of the reality of psi, myself among them, rarely think through all the implications of what is known about it. For example, the scientific evidence for precognition is as strong as that for telepathy—in fact some parapsychologists believe that most, if not all, ESP is a form of precognition. In addition to lab studies and the well-documented personal experiences of thousands of people, there were instances of precognition during the military Star Gate program in which remote viewers saw significant events before they occurred. Yet if this can happen, then the universe doesn't merely work differently from the way materialists believe, it works differently from the way any of us have believed. We are forced to realize that time doesn't flow in just one direction and the principle of cause and effect is false. Some quantum physicists (and many mystics) believe the past, present and future exist simultaneously. It's easy to imagine a world where people can communicate telepathically, but harder to picture one in which time as we know it is an illusion. Philosophers have done so, but it raises so many unanswerable questions about ourselves and our society that as a practical matter we have to ignore it except on the rare occasions when a premonition is experienced.

The more scientific knowledge a person has of the universe, the more upsetting it is to see evidence that reality doesn't follow the rules that have been thought to describe it. It might seem that physicists would be receptive to ideas that are counterintuitive, since they accept the theory of relativity despite its conflict with common sense. And they're aware that the underlying nature of things is, at the quantum level, based on probability rather than certainty—which is also true of psi. But in the case of physics, scientists have a defense against the disturbing nature of this realization. Most of them think about it only in terms of mathematics, and the more complex the math, the more effective it is as a distraction.

Although the study of particle physics depends on statistics, physicists tend to overlook that fact and claim that statistical studies of psi show effects too small to be significant, despite statements to the contrary by respected statisticians. Otherwise they would be forced to confront the fact that they can't understand what's going on. And that's something scientists aren't eager to do.

Anecdotal evidence—reports of individual human experiences with psi, regardless of their prevalence—is scorned by most critics as meaningless. Yet all scientific investigation begins with personal observation. In the "hard" sciences it is then possible to set up experiments with reproducible results. In other sciences, such as psychology, sociology, and even evolutionary theory, it often is not; scientists must judge theories in the basis of what they observe in the real world. The same should be true in parapsychology, since laboratory experiments, though statistically valid, cannot produce truly revelatory forms of psi that occur only when emotion is involved.

The reality of psi is not a matter of what can be proven. Either there is something in humans (and animals) that physics cannot ever explain, or there isn't. If there is, denying it won't make it go away, any more than believing in it will create it. Reductionism, which means reducing everything to the principles science has established and ignoring everything that doesn't fit, is at best a cop-out. We tend to forget that most of what we now know would never have been discovered under that philosophy. Most scientists of Newton's time resisted his ideas because they considered gravity an "occult" concept; nothing could work at a distance, they said. If gravity weren't impossible to ignore, they might have continued to think so.

Thus parapsychological research is not going to convince society that psi is real—the proof is already there insofar as statistics can provide it. And in fact, as soon as research starts to zero in on psi powers, the opposition may increase. When it begins to look even to mainstream psychologists as if psi might be real, people's unconscious fear of losing their basic orientation to the world may take over, so that more of them actively suppress any mention of it—not through censorship but through denial and ridicule. Researchers may stop going public, saying that such experimentation as had seemed successful had not panned out. Psi may be dismissed as an immature notion no longer taken seriously..

Yet the truth can't be buried forever. Eventually, the accumulation of psi experience by more and more individuals will convince the public of its existence. Most scientists, however, will continue to deny the evidence long after the reality of psi is demonstrated. They won't be convinced until there is a viable theory of how it works. That may be a long time in coming. But once such a theory is developed, scientific acceptance will grow.

Will there be a period during which people with psi powers are persecuted, as in my *Captain of Estel* novels? I'm afraid there may be. What humans fear, they often try to destroy; furthermore, minorities are most apt to be targeted when they seem likely to become a majority. A change as profound as widespread use of psi is bound to result in upheaval. But that will pass, as stages in evolution always do, and once most people are able to sense each others' feelings, there will be less misunderstanding among them.

I believe that it's a good thing that our planetary culture has so far focused on technology to the exclusion of psi. Advanced technology is essential for attaining interstellar travel, without which our species can't survive; and to devote too much of our minds to psi would be distracting. But over the long term, human progress requires development of both technology and mind powers—neither is sufficient without the other. Creating a technological civilization that incorporates psi will be the next step in our evolution. In my fiction I have said it's the criterion for contact with extraterrestrial civilizations, and I think that may be true.

Albert Einstein is widely quoted* as having said, "I have yet to meet a single person from our culture, no matter what his or her educational background, IQ, and specific training, who had powerful transpersonal experiences and continues to subscribe to the materialistic monism of Western science."* This, then, is the answer to the pervasive materialism that now hinders scientific advance. It cannot be overcome by persuasion, but as more and more people have such experiences, it will eventually be outgrown.

There is some doubt about the accuracy of this quote, as no source for it has been cited and the word "transpersonal" was not in common use during Einstein's lifetime.

Transhumanism Is a Dead End

I wrote parts of this essay in 2003 in an online forum where it met with a good deal of very emotional opposition. I have reorganized and expanded it to include some of my responses to the critics.

*

Transhumanism is the belief, now increasingly popular, that human beings can and should evolve beyond their biological physical and mental limitations through the use of science and technology. This includes enhancement of the body by means of genetic engineering or implants and of the mind by means of neural interfaces with computers, among other things. A prime aim of transhumanists is the extension of human lifespan, and most feel that the ultimate goal is immortality.

Some transhumanist proposals are personally repugnant to me, but that's not why I dislike the philosophy underlying them. In my opinion anybody who wants to become a cyborg, or to continue forever in the very limited state we know as life, has the right to make that choice if and when such options become available. The trouble with the transhumanist agenda lies not in what it might permit people to do in the future, but in how it leads them to perceive human nature. It is based on a narrow conception of mind that rules out vast areas of human experience; those that don't fit are simply shut out of its proponents' awareness. And that is hardly an attitude conducive to the advancement of science.

Transhumanism is based on a wholly materialistic conception of reality. In effect it says, "What we now understand about the mind and consciousness is all there is to it; there's nothing left but to fill in the details of how the brain works." It goes without saying that similar assumptions once made in the field of physics, such as Lord Kelvin's famous 1900 statement that "There is nothing new to be discovered in physics; all that remains is more and more precise measurement," proved to be short-sighted.

For millennia the majority of the human population has believed that the mind is more than a biological machine, that it extends in some inexplicable way beyond the individual body both in space (ESP) and in time (continued consciousness after death). The scientific evidence for the former can be denied only by those with closed minds; with regard to the latter, which is outside the present boundaries of science, it is well to remember the well-known aphorism, "Absence of evidence is not evidence of absence." The machine model of the human mind makes no provision for any data pertaining to what is increasingly coming to be referred to as "non-local mind." Such data is necessarily rejected by anyone bound by that model's premises, unconsciously if not consciously. And thus if that model is adopted even more widely than it is today, the progress of science—of unprejudiced investigation of all aspects of the universe whether or not they fit today's preconceptions—will be stalled.

Precisely because I am a strong supporter of science, I don't believe any area of reality should be arbitrarily ruled outside its field of inquiry. As readers of my novel *The Doors of the Universe* know, I am not opposed to genetic engineering of humans, and I think a ban on this or any other emerging application of science would be disastrous. I'm all for continued research in biotechnology, artificial intelligence, and whatever else transhumanists wish to devote themselves to. But not at the price of cutting off all other paths of research essential to our understanding of what it means to be human.

The term "transhumanism" is a misnomer for the developments it purports to describe, since the use of increasingly-advanced technology has always been a defining characteristic of our species. Becoming cyborgs will not make us less human, either in the transhumanists' sense of "beyond human" or in traditionalists' sense of degradation, any more than the advent of implant surgery or "test tube babies" altered our essential nature. To adapt ourselves and our environment to our needs through technology *is* our nature; it is what humans do. It is not all we do, however, and any philosophy that fails to acknowledge that fact is doomed to eventual obsolescence.

So I believe that transhumanism as a view of life, as opposed to its advocacy of specific technologies, is a dead end—as will become apparent when attempts to upload consciousness into a supercomputer fail to produce the desired result. I certainly don't oppose such attempts. On the contrary, I think that they will demonstrate beyond all doubt the naiveté of transhumanist assumptions about reality.

*

Robert Heinlein, whose science fiction was indisputably far-sighted and supportive of technology, was urged to agree to be cryogenically frozen after his death—a process transhumanists believe may lead to revival of the dead in the future. Alcor, the company that's doing this, was even willing to waive the usual fee for him. He repeatedly refused. It later came out that he had said to a friend, "How do I know it wouldn't interfere with reincarnation?" (Reported by Spider Robinson in *Requiem*, ed. Yoji Kondo, p. 408.) This doesn't mean that Heinlein believed in reincarnation in the sense that New Agers do; almost certainly he did not, and anyway, if some form of reincarnation does exist, it's unlikely that freezing bodies interferes with it any more than cremation does. The point is that he was unwilling to be identified with a reductionist view of mind that equates it with the physical brain and holds preservation of the brain to be in itself of value.

The goal of achieving immortality, either through physical modification of the body or through uploading a person's consciousness to a computer, appears to be a major attraction of transhumanism. It is a way for members of a society that no longer believes in any form of afterlife to cope with the universal fear of death. This no doubt accounts for the emotional fervor with which transhumanists defend an envisioned development that strikes many of us as both unlikely and unpleasant. Presumably its advocates have never been so unhappy as to feel never-ending consciousness would be a burden. If so they are to be envied, for whether or not one believes this life is all there is, a time comes. in old age if not sooner, when one perceives that eternity on Earth—in an enhanced biological body, an artificial one, or a supercomputer—would be either harrowing or intolerably boring. I, for one, can't imagine wanting to be uploaded; I'd rather take a chance on there being some form of existence a little less confining.

I do agree that a longer lifespan, if not burdened by the disabilities of old age, would be desirable—what dismays me is the idea of an endless one. I do not think even increased intelligence would produce the "infinitely interesting experience" some claim, because the amount of input data available to an embodied mind, whether biological or machine-based, would not be infinite.

Transhumanists are quick to say that people who found immortality boring could kill themselves. This is questionable; in a society that viewed perpetual life as an ultimate good, suicide might well be illegal, as it is in my novel *Stewards of the Flame*. A computer to which minds were uploaded would probably be programmed to prevent them from being deactivated; certainly it would if artificial intelligence, designed to protect human life at all costs, was involved. In any case, many people consider suicide wrong, if not on religious grounds then simply because it is a denial of any intrinsic purpose in life, a denial that would make all previous endurance of suffering pointless.

I have been told by a transhumanist that "Virtual worlds can be much more fun than this one," which is surely debatable—how many people would consider virtual sex as good as the real thing? Certainly most of us value physical contact with nature, the sensual enjoyment of sunshine and food, and the touch of loved ones. Though to the average person being permanently cut off from these would indeed seem confining, it has been argued that sex, and presumably all sensation, is only something that happens in the brain. So then perhaps in the transhumanists' ideal world no one would choose to have a body and no more babies would be born. Putting aside what would happen to the experiments with less-than-perfect results (for surely uploading would not have a 100% success rate when it was first offered to the public), what about the intermediate period when some people had bodies and other people didn't? Lovers, I should think, would have to arrange to be uploaded simultaneously or else experience a good deal of grief over the loss of contact with each other. Or is it assumed that love, as distinguished from sensation and exchange of thoughts, would also be considered obsolete?

The reason many find transhumanism objectionable is that unconsciously if not consciously, they perceive that the transhumanist agenda fails to take the "whole person" into account. And incidentally, the unconscious mind appears to be ignored by transhumanists and artificial intelligence enthusiasts alike. This is a case of going off half-cocked without consideration of a fundamental aspect of humanity that's known to exist but is not yet understood. It has been recognized for more than a century that a large part of the human mind is unconscious; how then could uploading consciousness into a computer produce the equivalent of a human being?

Objections to transhumanism's reductionist outlook are often assumed to be based on religion, as indeed some are. But by no means all opinions that don't fit prevailing scientific dogma are religious, and even they were, that would not invalidate the concepts underlying them. After all, at one time many things now understood by science were dealt with only by religion. Putting aside the beliefs of the ancients, as recently as the nineteenth century belief in life on other planets was widely accepted by both scientists and the clergy on the grounds that "God would not have made a useless world." Not only was there was no scientific evidence for the existence of exoplanets, let alone for life on them, but it was believed that such evidence would always be impossible to obtain. Today, when the question of whether extraterrestrial life exists is debated on scientific grounds, we see that past conceptions of the boundaries of science were too narrow. In my view, the goal of science is to understand the total range of universal reality, not to define reality in terms of its present capability for gathering evidence and dismiss what doesn't fit as inherently "religious" and therefore unworthy of acknowledgement.

I have no objection to the view that advanced technology will help to improve human capabilities, so long as that outlook does not close the door to scientific investigation of capabilities that are not physically based. I do not mean that mystic and/or metaphoric explanations of those capabilities should be accepted, and I'm aware that many scientists are turned away from that whole area of human experience by the huge amount of nonsense published that purports to explain it. But serious study of the so-called "paranormal" (which of course, if it exists, is not in any way supernatural) reveals that data concerning it cannot be simply swept under the rug on grounds that silly explanations for it have been offered.

I do object to equating the word "rationalism" with "materialism," as transhumanists are prone to do. Rationalism, according to the dictionary, is "a view that reason and experience rather than the nonrational are the fundamental criteria in the solution of problems." I contest not this principle but the unproven assumption that we cannot eventually explain non-material aspects of mind—which presently can be shown to exist through statistical methods, but not explained—by means of reason. Reason is wholly dependent on premises, from which it directly proceeds. Materialistic premises about the mind are basic to transhumanism and by limiting its perception of reality to them, it discourages the questioning of these premises by the people best qualified to investigate alternatives. Can we not develop technological means of extending human capabilities without adopting a worldview that narrows the potential scope of scientific discovery?

As to arguments against materialism that really are religious, transhumanists generally make false assumptions about their nature. Very few people other than fundamentalists (who despite their high profile in the media, are a minority among Christians) believe in a conventional heaven or hell. The afterlife is usually viewed as a state of being perhaps best defined, by analogy, as another dimension with different rules. It is felt that in this state of being there is access, perhaps infinite access, to knowledge that is not known, and in principle cannot be known, on Earth (or elsewhere in the universe as it is perceived in our present state of existence). Who does not wish that all the answers—not merely those a machine could find through analysis of existing data—could in due course be personally obtained?

The thing about machines is that they are built and the builders, even if previous generations of machines, necessarily proceed on the basis of premises. Yet in another state of being the very premises might be altered, thus permitting a level of knowledge that could never be attained in terms of existing ones. To be sure, it may be that there is no such state. If we bind our consciousness to machines or artificial bodies, we will never find out.

But I don't really believe that, because I suspect that if minds are not wholly material, they would not stay bound to machines. They would enter whatever other "dimension" exists anyway when ready to do so. Thus I am not any more worried about being trapped forever in a machine than I am about becoming a spirit walking around in chains like a ghost in a horror story. Like the characters in *Stewards of the Flame* who overcome their fear of permanent stasis, I know underneath that the truly horrible thing is not the attempt to preserve life after it's gone, but the effect that preservation of pseudo-life has on the living.

Throughout human history, in all cultures, there has been a prevalent belief in the existence of an aspect of the mind or soul apart from the physical body, and while collective belief doesn't prove an idea is true, it does indicate that what's behind the idea should sooner or later be investigated. In all probability "where there's smoke there's fire." Science often proves a widely-believed explanation of a phenomenon to be wrong—and this is as true of scientific explanations as of earlier non-scientific ones—but it rarely if ever finds that there was nothing to be explained.

Of course an afterlife, if any, is hardly the main issue involving this other aspect of reality. The so-called "paranormal" abilities of humans while alive are what matter to science. Most explanations people have had for such things are, in my opinion, metaphors—metaphors being the means human minds employ to deal with things they don't understand. I do not think any of these metaphors will turn out to be literally true, any more than the idea that disease is caused by evil spirits is true. The question is, what's the foundation of the metaphors? We have no evidence whatsoever for saying there isn't one.

The "biological machine" model of the mind is also a metaphor. It's useful for practical scientific work; metaphors exist precisely because they are useful. But like all metaphors, it is based on unverified belief, as is every scientific conception of something not specifically proven or disproven. There is no justification for treating it as indisputable fact.

Historically, science has progressed not only by learning more and more in the specific fields it first investigated, such as astronomy, physics, and medicine, but by gradually expanding its scope into fields less easily investigated. Molecular biology and cybernetics are among the most recent examples. Scientific work in psychology has barely gotten off the ground; theories in that field have been primitive and inaccurate, but it's recognized as a legitimate field of study. Social sciences are not yet very scientific, though the advent of complexity theory will, I think, bring about progress in them. But parapsychology is struggling; only a few scientists are courageous enough to buck the opposition and devote their working lives to it.

Lack of funding and disrespect are formidable barriers to progress in any field. If transhumanism as a philosophy, with its commitment to a focus on material "perfection" of the human race, dominates the twenty-first century, then increased funding, and therefore increased respect for parapsychology, is not likely to materialize; thus it will be left by default to people unqualified to deal with its data scientifically. I want to see science widen its knowledge of reality, not shrink back within old boundaries. To ignore evidence for phenomena such as psi capabilities that reductionistic premises cannot explain—to pretend this evidence doesn't exist in order to bolster the belief that we're somewhere near to understanding the mind and can ultimately reproduce it via technology—is to set a goal for science, and for humankind's future, that is not high enough.

Like the burgeoning field of artificial intelligence, trans-humanism offers premature answers to long-standing questions that have been debated since ancient times. It claims that the debate is over, and thus seems to be leading scientific inquiry into a dead end. If it were to succeed in its major aims a crucial aspect of human existence would remain unexamined and unacknowledged.

But I don't think that is going to happen. Rather, I believe—as I have believed for many years—that when an artificial or artificially-embodied mind technically indistinguishable from a human mind is finally produced, the difference between the two will be so apparent as to invalidate the restrictive premises under which it was created. So perhaps transhumanism is not a dead end after all, if failure of its prime goal proves to be the means by which those premises are overturned

The Worship of Medical Authority

This is a portion of the introductory material I wrote in 1995 for an online course I was teaching through Connected Education for credit from the New School for Social Research. Titled "Science and Twenty-First Century Medicine," it was a Media Studies course, so the focus was on public attitudes toward medical advances. I have made minor wording alterations in a few places, added several paragraphs, and inserted a few references to my novel Stewards of the Flame, *which was written ten years later. Otherwise it needed no revision; the public's view of medicine has not changed.*

*

Why does our society view the medical establishment as the supreme authority on what our priorities should be and how we should live our lives? Hope for cure of all illness, based on medicine's clear success in dealing with injuries and infectious disease, is understandable. But there is far more to the public's faith in it than that. It's more than mere wishful thinking, too, although that certainly contributes.

I suspect one factor is that we're conditioned from infancy to rely on authority figures for our well-being—conditioned by the fact that we're a species with a long childhood, during which care from adults is essential to survival. In former centuries, experience with medicine was less apt than it is today to inspire trust in doctors as substitute all-knowing parents. But physicians, who were once the social equals of their patients, assumed an authoritative role with the advent of hospital medicine.

There is, to be sure, a present trend toward rejection of doctors' authority. Many observers have remarked on this as a distinct change from the attitudes that prevailed in the first half of the twentieth century. However, it applies mainly to the authority of individual doctors, in the sense that they're no longer placed on pedestals and are often suspected of incompetence or financial greed. People are now quick to seek advice from a sequence of doctors and/or alternative practitioners when dissatisfied with the help they get. Yet the belief that medical providers of one sort or another *can* help, that they're the ultimate source of advice to which humans have access, seems stronger than ever.

Personally, I believe that neither medicine's past success nor conditioned dependence on its authority is enough to account for the fanatic devotion our society has to the concept of medical care. It has always seemed to me that such devotion is comparable to a religion. Some years ago, when I began to investigate these issues, I imagined I might write an essay setting forth that hypothesis—though I knew it would be regarded as heresy. But I soon found that it's far from an original idea. Many writers have argued that medicine has become a religion; the more one examines the analogy, the more obvious it appears to be.

As faith in traditional religions declined, and people stopped turning to the clergy with their problems, they turned to doctors and psychotherapists; this much is common knowledge. Less widely recognized is the fact that instead of the secularization of society that's generally assumed, we got medicalization. Our culture is not without a creed. Its creed is a list of questionable assumptions about the scientific validity of modern medicine, plus some others that are not related to science.

Chief among these other assumptions is the *moral* authority accorded to medicine. In our society, it's not just thought wise to defer to medical judgment—it's considered virtuous. Dissent on the part of a patient, or worse, the refusal to become a patient, is an offense against common mores; the admonishment "You ought to see a doctor" implies reproach, and rebellion is seen as scandalous. Failure to follow medically-endorsed health recommendations is viewed as sin; it has reached the point where ordinary foods once found on everyone's table are referred to in recipe columns as "sinful." The primary emotion now felt by people in connection with their health is not satisfaction or dissatisfaction, but guilt. A person who values quality of life over hypothetical length of life is looked upon not merely as foolish, but as irresponsible. Acceptance of unpleasant treatments or lifestyles, whether or not these can rationally be expected to impart benefit, is considered commendable, while rejection of them is regarded as error—as if, perhaps, doing penance for ill health might exorcise it, or at least ward society against its spread.

It is, of course, an ancient belief that sickness is punishment for sin. But today, that belief has been explicitly rejected, to be replaced by a feeling that poor health is in itself a moral failing. We strive not to be good, but to be well. Physician Arthur Barsky, in *Worried Sick*, writes, "As people in the past sought to lead a 'religious life,' we now seek the healthy lifestyle. Attaining a sense of wellness is like attaining a state of grace. In a sense, we've replaced the religious quest for the salvation of our souls with the secular quest for the salvation of our bodies." Quoting Marshall Becker, he continues, "Health promotion . . . is a new religion, in which we worship ourselves, attribute good health to our devoutness, and view illness as just punishment for those who have not yet seen the way."

There is irony in the fact that some criticize our society for abandoning the concept of moral responsibility and replacing it with the idea that offenders are "sick," for in fact, sickness has indeed become the equivalent of sinfulness—and treatment the equivalent of repentance—not merely in the case of criminals, but with regard to all who display physical or behavioral deviance from an ideal "norm." The sick person is required to submit to the authority of the establishment, just as sinners once were. His or her private judgments are not accorded any more respect than they were by the most dogmatic of churches. To be sure, treatment, except for certain "mental illnesses," is not quite compulsory (yet) unless a crime is committed; we don't subject patients to force, though we do sometimes induce them to undergo torture for the sake of alleged salvation. But the source of pressure to conform is identical: preserve society's faith at all costs! We may say our society is faithless, that it has lost its sense of common purpose; but that's not true. We worship Perfect Health and the medical establishment is its priesthood.

None of this is meant to suggest insincerity on the part of doctors. Most doctors are motivated by genuine desire to help their fellow human beings. But so were most priests, even within intolerant religious establishments. They sincerely believed that sinners would be saved by deference to ecclesiastical authority, and were sure that nonconformists would end up in hell. Similarly, doctors believe that people's ultimate welfare depends on adherence to the dictates of health dogma and that disregard of these will result in suffering or early death. They trust the medical science they have been taught, as clergy trusted the theological hierarchy; they're convinced of the efficacy of its rituals. Significantly, this conviction comes from faith—fewer such rituals than we suppose are backed by objective science.

It's not really surprising that this faith has developed. Early death was once common, so common that people didn't question it—but they had the hope of heaven. Most of them paid a great deal of attention to doing what they were told would assure them of heaven. But belief in a literal heaven has faded. To today's majority, religious rituals, if meaningful at all, have symbolic rather than practical value. Is it any wonder that in an era where medicine can defy death, the goal of maximum lifespan—pursued by often-slavish devotion to ritualistic health practices—has taken heaven's place?

People have always sought direction in their search for salvation and have been willing to give up a lot for it. Thus they've welcomed the medicalization of our society, assuming the role of patients even when they don't feel sick and going for checkups as they once would have gone to confession. Craving absolution, they now subject themselves to medical authority not just during illness, but at birth, death, and throughout all phases of their lives.

Nor is that the only way in which medicine assumes the role of religion. To many, it is the prime source of mystery and awe. Once, people sought cures at shrines and temples, as some do even today; to look to an arcane source for healing is a deep-seated human impulse. Anthropologist/physician Melvin Konner writes that in many cultures doctors still keep secrets from their patients, as they used to in our own, and that only in theory does modern medicine's focus on science dispel, rather than cultivate, mystery. "In reality the mystery remains, but it has a different location: it lives on the frontiers of technology . . . doctors and patients stand together, grateful and humbled, before what almost seem to be technological gods." In their book *The Healing Brain* Robert Ornstein and David Sobel quote a description by psychiatrist Jerome Frank of "how a great teaching hospital might look to an anthropologist from Mars who is studying healing shrines in America." This beautifully illustrates the medical/religious analogy.

I suspect that the idolization of medicine and its technology is a trend that won't change in the twenty-first century, if it ever does, though many assumptions of today's medical establishment will be altered. Personally, I tend to get emotional about it, as heretics often do, and my dystopian science fiction novel *Stewards of the Flame* deals with a society in which this trend has been carried to its logical conclusion, in some respects to reductio ad absurdum lengths. Yet I don't mean to say that it's bad in all ways. As Ornstein and Sobel point out, it may serve to mobilize "the faith that heals" in those who don't share my skepticism.

My point in raising the issue is simply to explain how we got into a situation where public perceptions conflict with an objective approach to the findings of science. There has been, and still is, a lot of resistance to those findings, especially in the case of evidence that medicine can't magically perfect us. In my opinion, the religious character of our medical outlook is a principal reason for it.

*

Just as political and financial factors once shaped the doctrine of authoritative legally-established religions, they have profound effects on the dogma promulgated by health authorities today. Our medical-industrial establishment (a term more accurate than mere "medical establishment," if more clumsy) is exceeded in size only by our military-industrial establishment, and it is no more efficient or high-minded. It encompasses powerful commercial interests and a huge government bureaucracy. And although as a general rule, I believe the profit motive leads to benefits for the public, I do not think this is true with medicine. In the case of other technologies, we can judge as individuals how much we'll gain from them; we can buy or not buy a product, as we choose. Not so with medical technology. We are told what we need not just by advertising we can tune out, but by authorities we're told to trust. We're in no position to evaluate their pronouncements, let alone associated media hype. A good deal is now said about how commercialization of health care has led to corners being cut in provision of services, but the opposite problem is older and more pervasive.

Medical research is expensive. A lot of it is paid for, directly or indirectly, by pharmaceutical firms; the rest is funded by government grants and/or by organizations that appeal directly to the public, and thus have an interest in public perceptions. The results of such research aren't of mere scientific interest; they impact the economy—suppliers, laboratories, and even the food industry, in addition to medical providers and the drug companies. They influence the prestige and power structure of institutions. In some cases, research projects affect local interests to the extent that politicians lobby for them. The insurance industry, despite reluctance to pay for specific services, nevertheless benefits from a situation in which health insurance is a necessity of life. Furthermore, the amount spent yearly on medical testing now exceeds that spent on pharmaceuticals, and much of it is paid on behalf of people who are not even sick. It's hardly surprising if from all this, a policy of maximum consumption of "care" emerges, and medical interventionism prevails over any suggestion that we need not all be constant consumers of it. People want "only the best" when it comes to health care, and are led to assume, not necessarily wisely, that "best" means "most."

The influence of pharmaceutical companies on health care standards is too well known, and too widely criticized, for me to go into here—there are countless books directed to the public that present the details. Briefly, constant lobbying results in legislation that keeps drug prices artificially high; the results of tests for drug safety and efficacy are not always as valid as they appear; scandal involving government regulation of medical drugs is by no means rare; new expensive drugs are aggressively marketed as "improved" replacements for older ones that work just as well if not better; and worst of all, drugs are widely promoted for conditions that don't require medication and in many cases would not even be considered "illness" if no such promotion had been done. To be sure, some of the immense profit derived from selling such drugs funds the development of those needed by people who are really sick. But at what cost to society? Since drugs are not profitable unless taken by large numbers of people, the result has been that a great many—and their doctors—have been led to believe that they need ongoing medication, to prevent future illness if for no other reason. They have also been conditioned from childhood onward to think that any form of discomfort should be dealt with by ingesting chemicals and that doing so is virtuous; is it any wonder that so many abuse drugs, or turn to illegal ones for relief from mere unhappiness or boredom?

I suspect, too, that there are additional reasons for government promotion of so-called preventative medicine. If it leads to only a minor statistical decrease in "risk factor" diseases—or can be made to seem as if it will—such propaganda is politically advantageous, even if it harms more individuals than it helps (which is more often the case than is generally known). The media attention to "prevention" makes it look as if the government is Doing Something about diseases for which cures haven't appeared. Moreover, the concept sounds good; it even sounds like the cost-saver it's asserted to be (although the General Accounting Office determined in 1987 that "preventive" medicine costs more in the long run than treating only people who get sick, something insurance companies have always known). Perhaps the bureaucrats have been taken in by the way it sounds. On the other hand . . .

Government involvement in medical care has been on the rise ever since public funds were first devoted to it, and has increased exponentially since the establishment of Medicare and Medicaid. Further intrusion on personal health autonomy can be expected if coverage of medical care costs is broadened—what the government pays for, it controls, and to bureaucrats, any opportunity to control is welcome. Surprisingly few warnings about the consequences are coming from people who ordinarily are deeply concerned about protection of individual liberty; on the contrary, virtually any extension of government authority is viewed with approval if it is deemed justifiable as a health measure. I think the situation I portrayed in *Stewards of the Flame* is not too far-fetched.

I suppose all this makes me sound like a cynic—which, in other areas of inquiry, I usually am not. I ask myself, sometimes, why I have never shared society's trust in medical authority, why, even as a child, I reacted strongly against well-meant attempts to interfere with people's bodies. It struck me as an unwarranted invasion of privacy. The purpose of medicine, I felt, was to relieve suffering, not to concern itself with "norms." Perhaps it was simply that I value individuality over conformity. Or perhaps it was that by nature I'm an intellectual rebel, inclined to doubt anything I'm old I should believe. Many people react that way to religion, but I was never urged to accept a formal religion; perhaps for me too, in a negative sense, the medical establishment took its place.

*

Although commercial interests have a powerful role in determining the perceived status of an individual's health, the influence of social factors is even stronger. Like any other religion, medicine defines the need for reformation in terms of deviance from what is currently seen as normal.

In *The Limits of Medicine* Edward Golub points out that "our perception of when we are well or ill is defined by time and culture." He suggests that the gnarled hands of arthritis were once viewed not as disease, but as the natural condition of those lucky enough to live to old age. In general, however, atypical conditions are not seen in a positive way. Rather, variations from idealized "norms" are arbitrarily declared to be signs of sickness. Furthermore, as was observed by a medical news publication, "Politicians discovered long ago that a good way to pass the buck is to get a problem labeled as a disease." The medical establishment, as the agent of society, is given sole power not only to cure illness, but to define it—and the definitions, upheld by its authority, both reflect and reinforce society's prejudices. Consider these examples:

1. In the 1850s the desire of slaves to escape from their masters was labeled a mental disease, at least in the South, and was given the scientific-sounding name of "drapteomania."

2. Throughout the second half of the nineteenth century, the normal emotional problems of women were blamed on their reproductive organs and were frequently treated by gynecological surgery.

3. In the early twentieth century, identification of "undesirable" traits as disease appropriate for extirpation by physicians was not limited to Nazis. The eugenics movement in America resulted in

20,000 involuntary sterilizations, and was lauded by respected authorities as an advance in medical science. Though focused on the "feebleminded," it also targeted "drug fiends" and "drunkards"; even "shiftlessness" and "pauperism" were assumed to be medical conditions of genetic origin.

4. Frontal lobotomy was performed on thousands of "disturbed" patients, even children, in the 1940s despite growing criticism of its devastating effects. So great was its popularity that its inventor shared the 1949 Nobel Prize in Medicine.

5. From the 1920s through the 1950s, intact tonsils were viewed as unhealthy and tonsillectomy for children was virtually routine among families who could afford it; it was said that this would prevent colds. I myself was an unwilling victim of this practice in 1940, although a controlled study as early as the 1930s showed that children without tonsils have no fewer colds than others. The fact that colds aren't serious enough to justify surgery, which seemed apparent to me at the age of six, was not recognized during that era.

6. Until the late 1950s, enlarged thymus glands found in infants were called "status thymicolymphaticus" and treated with irradiation. Eventually, such thymus glands were recognized as normal—but children so treated had an increased risk of thyroid cancer and the girls had an increased risk of breast cancer as well. This is only one example of instances in which an unwarranted "disease" label has resulted in unalloyed physical harm.

7. Homosexuality was listed as a disorder by the American Psychiatric Association until 1973; only the gay rights movement brought about a change in its designation.

8. During the second half of the twentieth century the natural event of menopause was medicalized to the extent that some spoke of older women's loss of estrogen as a "deficiency disease"; doctors advocated hormone replacement therapy even while conceding that the "deficiency" is universal. Then in 2002 a major study revealed that hormone replacement leads to an increased risk of cancer and heart disease. Today it is generally advised only for women with abnormally serious menopausal symptoms.

9. Obesity is currently defined as a disease and is often aggressively treated, despite lack of any proof that fatness per se is harmful. There are statistical *associations* between fatness and some diseases, but in most cases no evidence that it is a causal factor. Furthermore, there are associations between fatness and *decreased* risk for other diseases.

10. There is considerable concern among critics of gene therapy about the possibility that other ways in which individuals naturally differ, or variations leading to minor impairments and/or mere risks, will be erroneously perceived as health defects once genes associated with them can be identified. This has already happened in the case of some carriers of recessive genes.

11. Biochemical treatment of children who are not sick, but merely atypical, is already common. For example, growth hormone produced through genetic engineering is given to those who are short, and amphetamines are routinely prescribed for the "overactive." Both these practices have drawn heavy criticism, but they're on the increase and further biotechnical developments will lead to additional ones.

12. Conditions identified as alleged risk factors for disease are increasingly viewed as diseases in themselves and are believed to require treatment, often at the cost of quality of life, despite the fact that most individuals with these conditions don't die at a significantly younger age than average. The mere potential for illness (which exists, in one form or another, in all of us) is thus no longer distinguished from real illness. It remains to be seen whether the routine prescription of statins to lower cholesterol will prove to be one of medicine's tragic mistakes; giving powerful drugs to a large percentage of the population without knowledge of their long-term effects strikes some people as a prescription for disaster.

13. Today, normal emotions such as anxiety and depression are considered illnesses even in people who are not incapacitated by them, and are given ominous-sounding diagnostic designations. Use of drugs to suppress them is not only condoned, but promoted as virtuous; people who refuse to "get help" are looked

down upon. In general, unhappy or withdrawn people are viewed as less healthy than "well-adjusted" ones, whether or not they display physical symptoms.

14. Addictions are now generally considered diseases; in fact, to question this label is to be thought moralistic and unenlightened— although the most effective ways of dealing with them aren't of medical origin. Increasingly, co-dependency is also looked upon as sickness, for one reason because, in one observer's words, that creates "a whole new class of billable patients."

15. Antisocial and/or deviant behavior has always been branded "mental illness"; the extent to which this is a valid concept is highly controversial. In the former Soviet Union even dissident opinion was viewed as sickness, while at the other extreme, psychiatrist Thomas Szasz argues that there is no such thing as a real mental illness and that schizophrenics are merely victims of political oppression—a view strongly endorsed by the "Mad Rights" movement among mental patients who feel that as long as they aren't harming anyone they should not be considered ill. The position now prevailing among psychiatrists is that atypical mental functioning is of biochemical origin; increasingly, it is treated with drugs that some experts call "chemical lobotomy."

16. It is presently unacceptable to list "old age" as the cause of death on a legal death certificate; doctors are required to pinpoint a specific disease. Some feel this is unrealistic, since organs normally fail in old age, often at about the same time. The result is first, that "cures" for such "disease" are expected, and second, that even a person who has lived a long and healthy life can't avoid "succumbing to disease" at its end.

What all these examples have in common is the fact that they involve people who would never consider themselves sick and in need of medical care if the medical establishment, and/or society, did not so decree. I do not mean to suggest that people in those categories don't ever need advice and support (though to call helpful counseling "medical" is at best a means of getting medical insurance to pay for it—a benefit that may or may not outweigh the psychological cost of being viewed as diseased). My point is that all such "sickness" labels are value judgments, and depend on attitudes toward human diversity. Medicine has been assumed to know what's pathological as distinguished from what's merely variable, but that isn't something science can determine.

We might say that the ability of a person to function adequately would be a reasonable criterion. But even that depends on culture; is nearsightedness pathological, when in our era, many of us get along fine with glasses? Increasingly, the disabled have been protesting against the idea that they are defective, and are demanding cultural accommodations that will facilitate their functioning as full members of society. Thus the problem, perhaps, lies not in the definition of "defective," but in the very concept of a defect; and this problem will grow as more and more is learned about the human genome.

There is, to be sure, a common-sense definition of illness on which we can all agree as a minimum: it's the state where a person is truly suffering—not just assumed to be suffering—and/or is unable to engage in his or her regular activities (and where the problem couldn't be solved by elimination of social prejudice). If this were the criterion for medicine's involvement, there would be little difficulty in determining its purview. But, in our medicalized age, medicine's aspirations go far beyond the relief of suffering and incapacity. As indicated above, they extend to the enforcement social conformity.

Although intolerance toward minorities has supposedly become taboo in our society, in the case of health-related issues the opposite trend is accelerating. Mike Fitzpatrick, in a 2004 article in *The British Journal of General Practice*, points out, "Resistance to medicalisation was a common theme of the anti-psychiatry, gay liberation and feminist movements of the 1970s. But [now] ... radical campaigners are more likely to demand medical recognition of conditions such as ME or Gulf War Syndrome than oppose it as pathologising and stigmatising. Whereas, in the past, the pressure for medicalisation came from the medical profession or the government, now it also comes from below, from society itself."

Furthermore, the larger aim of medicine—strongly supported by the public—is now to *prevent* illness; in pursuit of this aim, it brands individuals believed to be "at risk" as already ill, often subjecting them to far more suffering, both physical and mental, than they would otherwise experience, not to mention the ongoing financial burden and inconvenience. "Early detection" is encouraged, though it often uncovers conditions that might never have developed unpleasant symptoms. Is a symptomless illness really an illness? Do factors that may or may not contribute to one's eventual death qualify as disease? It is society, not science, that says so; after all, we will all die of something, sometime, and all physical predispositions may eventually be detectable.

Given the fact that ideas about health are indeed social rather than derivable from objective science, we may ask if there is any such thing as "perfect health." Obviously, there is not. Nor is there any such thing as "fitness," another concept widely abused today—too seldom do we ask, "fitness for what?" Fitness depends on environment. It's meaningless to say that most of us aren't "fit enough" to survive in the environment our ancestors did, since that's not the environment we live in. Then too, as the term is used in evolutionary theory, fitness means reproductive fitness: not ability to survive to old age, but the ability to get genes into succeeding generations. In our culture, this hasn't much correlation with illness or absence thereof in adulthood. We may associate general fitness with health, but it's circular to define them in terms of each other. Nor does it make sense to define health as the physical condition characterizing humans of prehistoric times, when not only were lifestyles unlike ours, but hardly anyone lived past thirty.

Furthermore, many of the factors that we assume constitute health are self-contradictory. As is explained in Randolph Nesse and George Williams' book *How We Get Sick*, evolution leads to trade-offs. Some apparent illnesses are actually the body's defenses against worse illnesses. Some are the result of genes that also have beneficial effects. In the psychological realm, anxiety is an indispensable part of the ability to avoid danger. We can no more say that there is "something wrong" with people for being sick or unhappy sometimes than we can say there's something wrong with members of racial minorities.

So if health isn't an objective concept—if there is no

standard by which some people can be called normal and others abnormal, and if sickness is inevitable and, in some situations, even adaptive—why do we ask medicine to make us perfect? This is an impossible goal, one leading to intolerance as well as a continuous feeling of guilt on the part of everyone who accepts medicine's authority. No matter how advanced our science becomes, it cannot be attained. Thus, there are limits to what health care can accomplish, and they are not determined by technological skill, let alone compliance with medically-advised behavior.

Yet as scientific knowledge increases, leading to the availability of tests that detect more and more about individual biochemistry, the tendency to blame people for their physical condition will grow. It will be said, as it often is already, that they should "take responsibility for their own health," meaning not that they should be willing to abide by the consequences of their personal choices, but that they should adhere to the pronouncements of the medical establishment—or worse, the government—about how they ought to behave. Once illness was viewed as misfortune, and before that, as punishment for sin; now the pendulum is swinging back and it's being attributed to sin again. As always, dogmatic authorities stand ready to absolve those who repent and accept the penance prescribed. Whether or not they do, if their genetic constitution happens to be such that statistically-based advice would be damaging to them, they will be out of luck. Nonconformists, whatever their reason for resistance, will be persecuted and, quite possibly, taxed.

But change may be on the horizon. Ironically in view of the historic trend toward medicalization, criticism of current practices is beginning to be voiced—not by the public, but by medical professionals. At the time my novel *Stewards of the Flame* was written, the belief that many standard medical tests and treatments are unnecessary or even harmful was rare enough to be called heresy, and very little was published expressing that point of view. Since then, increased attention has been paid to the harm caused by medical overtreatment and a number of books and articles about it have appeared, many of them by doctors.

Moreover, it has been pointed out that a great deal of unnecessary testing and treatment is done because patients demand it, sometimes against their physician's best judgment. Although it is still common for doctors and hospitals to urge tests, procedures, and drugs merely because they are conventionally, if erroneously, viewed as beneficial—and in some cases, because they are financially profitable—the fundamental problem is the outlook of society as a whole rather than that of the medical establishment. Hopefully, the escalating cost of medical care will wake people up if nothing else does. Society will never be able to afford treatment for all those who need it as long as so much is spent on unnecessary and sometimes-harmful care for those who do not.

Why I Don't Read Much Science Fiction

Despite my interest in the future I have never been an avid science fiction fan, and it's hard for me to find SF novels that I enjoy. Here are some of the things that turn me away.

*

Strange as it may seem, although I have written eleven science fiction novels I'm not a science fiction fan in the sense that term is generally used. I actually haven't read much science fiction, certainly not recently. And the reason I haven't is that my tastes don't match those of most people in the field. This is also why my own novels have never been acceptable to mass-market science fiction publishers—I write the kind of book I like to read, as most authors do, and I don't like to read the same kind of story as the majority of fans.

One reason is that I'm not really interested in action scenes such as space battles, nor do I care about the technical details of future science. Another is that I don't enjoy the kind of science fiction that aims to get as far from real life as possible just for the sake of difference. I prefer stories focused on the concerns and feelings of human beings—and on what we may have in common with beings of other worlds—rather than wild imagnings about alien species wholly unlike anything we know. The biggest problem I have in finding science fiction I like, however, is that so much of it is based on themes that conflict with my conception of the universe.

I can put up with a lot that would cause some people to turn away from a novel. I don't mind if the science isn't accurate—for example, starships that defy the laws of physics as we know them, or details of technology wrong where only a reader knowledgeable about the subject would notice. It's okay if the characters speak and act like people of today; though we know our culture will change; we don't know how it will change, so it can't be portrayed realistically in any case. Small anachronisms sometimes do jar me, but I can overlook them if the main theme of the book is valid. And though I dislike graphic violence, explicit sex, and foul language, I can tolerate them where they are germane to the story rather than gratuitous.

But there are some things that rule out a book if I'm choosing something to read for entertainment—themes that are so far removed from my underlying beliefs about the human condition that they destroy any pleasure I might take in the story. Unfortunately, they are very common themes, which leaves me with little to read.

Before listing these themes, I should make plain that I'm not talking about literary quality here. There are novels of high quality based on many of them, and I'm not saying these don't have value. It does disturb me when an author uses such themes merely to sell books rather than from honest conviction, but whether that has been done in a particular book is not for me to judge. I'm simply explaining that for recreational reading I choose books I hope to enjoy, and I don't enjoy fiction that conflicts with my beliefs about the nature of humans and our relation to the universe. That said, here are the theme I avoid.

1. War with aliens. This projection of the past into the future is wholly anachronistic. Any aliens technologically advanced enough to undertake an interstellar war or invasion will have matured far beyond the stage of wanting to do so, just as we ourselves will have. Though there might be individual power-seekers among them, they'd be unable to attract a following. And aliens with starships will have no need to fight over resources, as in space there are plenty to go around. For more detail about my thoughts on this, see my essay "Why There Will Never Be an Interstellar War."

2. Aliens who play God. I can't give a definitive reason why I don't believe advanced aliens take it upon themselves to teach or admonish less highly evolved ones, since after all, we don't really know anything about the psychology of aliens. But this has been one of my core beliefs since as far back as I can remember, as readers of my novels know. I wouldn't go so far as to reject films like *2001*, which had enough more in it that I liked to compensate. I found *2010* harder to swallow. And I disliked episodes of Star Trek in which the nominal non-interfere policy seemed to exist only so that Captain Kirk could violate it. (Incidentally, the idea that aliens haven't arrived on Earth because of such a policy

on their part is now sometimes called the Star Trek hypothesis and I am sometimes assumed to have derived it from Star Trek, which irritates me since I developed that policy in an early draft of *Enchantress from the Stars* back in 1957.)

3. Aliens with grotesque shapes that couldn't possibly give them the capabilities they are presumed to have. Of course aliens aren't just like us. Personally I prefer to think of them as more less human in form because we don't know what they actually look like and if they're going to be portrayed unrealistically it might as well be in a way that makes it easy for readers to identify with the characters. But I can accept other conceptions of them as long as their anatomy matches their actions. If it doesn't, the story is just as fantastic as children's stories about talking animals.

4. Humans viewed as inherently inferior to aliens (as distinguished from less advanced) or as a danger to the galaxy that should be eliminated. This applies both ways—no aliens are inherently inferior either, nor are some innately more aggressive than others. There aren't any inferior races of intelligent beings, only immature ones. Hopefully the very concept of racism will die out in our culture long before it could affect interstellar relationships, but I don't think it's good to perpetuate it by applying it either to aliens or to ourselves as a species.

5. Humans trying to take over inhabited alien worlds. A story in which specific villains do so is okay, as long as it isn't implied that it's endorsed by Earth's authorities or that it's "human nature" to steal territory from others. Again, this is an anachronistic idea. The fact that it happened on Earth in past centuries does not mean that it could happen as far in the future as when we have starships—no authority on Earth would condone it even now. (Yes, a starfaring race does it in my own novel *Enchantress from the Stars*, but that is intentionally based on 20th-century mythology about space explorers, just as the portion of the book dealing with dragon-slaying is based on medieval mythology; it's not meant to be a realistic view of the future.) Human evolution involves progress on all fronts. We don't behave like our remote

ancestors and our descendants won't be behave as we did in the relatively recent past.

6. Conscious AI. This theme appears more and more often and I cringe when I encounter it, as it's the antithesis of everything I believe about the nature of human (or alien) beings. There seems to be a strong psychological tendency for people to personify inanimate things, which I have never shared. Now people even personify computers. I love computers and couldn't live without mine, but I turned Cortana off the day I got it. Even a semblance of conscious AI depends on the supposition that minds are nothing more than physical brains—the fundamental tenet of materialism. See my essays "Robots Will Never Replace Humans" and "The Roots of Disbelief in Human Mind Powers."

7. Human consciousness uploaded to computers or, even more impossibly, electronically transmitted to other worlds. This notion reduces minds to mere software. It would be laughable if it were not that some scientists take it seriously. The transporter popularized by Star Trek was imagined so that the producers wouldn't have to spend money on sets showing landers, which was a reasonable excuse for the acceptance of such a concept. I can think of no other.

8. Cyborgs shown as normal and/or as superior to humans. Neurotechnology is a good thing when it restores normal functioning to disabled people, but I find the idea of altering our bodies' natural design dehumanizing and even dangerous. Just as the science of genomics is discovering that individual differences are far more complex and less easily modified than we thought, future science may learn that there are good reasons why evolution made us the way we are.

9. Physical (or electronic) immortality. If such a thing were possible, it would invalidate everything we know about life and about evolution. This is a matter not of technology, but of basic concepts of what it means to be alive. Change, not stasis,

underlies all aspects of nature. See my essay "In Defense of Natural Death."

10. Psi powers viewed as unique to a particular alien species, or as a freak mutation. If such powers exist, as I believe they do in both humans and animals, then they are latent in all sentient beings. Some individuals are more talented than others in using them, but they aren't just a chance genetic characteristic like skin color. They are a fundamental aspect of the interface between the material and nonmaterial. The degree of their development varies; in my opinion, conscious control of them increases as evolution progresses.

11. Advanced aliens who have intentionally abandoned technology. Considering the technological developments I've listed as unacceptable to me, it might be thought that I'm opposed to high technology in general. I'm not. Technology and mind powers are equally essential to the advancement and survival of any alien species, as they are to ours. A species that abandoned a significant portion of its technology would not only stop evolving but would lose the accumulated knowledge that enabled it to advance, and sooner or later it would die out. A story about the process of abandoning it, or the loss of it through some disaster, would be a tragedy; and I'm rarely in the mood for tragedy.

12. Advanced aliens who don't have space travel. If advanced means more advanced than we are right now, this in an impossibility. No technologically-progressive species can survive indefinitely confined to a single planet. If it tried to, it would either die out relatively quickly or, if it managed through totalitarian government to control its population and use of resources, it would eventually explode into violence and destroy itself along with its environment. In neither scenario could it be considered "advanced" very long.

13. An implication that humankind is sinful because the environment on Earth is changing. In the first place, I believe more changes are due to other natural processes than is generally

recognized; we are not as powerful as we think. But where human activity does contribute to environmental change, it's because we do what is natural for our species to do: from prehistoric times onward, we have survived by altering our environment though technology. That is what caused us to become an exceptionally successful species in evolutionary terms, and in no sense are we blameworthy. All environments change. When ours changes for the worse it's a sign not that we are at fault for acting naturally, but that we have outgrown it and, like other species in that situation, we need to expand to a new ecological niche—in our case, space. Species that fail to adapt to changing environments or expand to new ones die out; that is a basic principle of evolution, and I believe it applies to alien worlds as well as our own.

14. A pessimistic or misanthropic view of humankind's long-term future. I don't like doomsaying and I don't think it's realistic. Barring some apocalyptic disaster, I believe we will continue to evolve and progress. The only disaster stories I like are those that focus on the indestructible quality of the human spirit and the conviction that we will recover from whatever losses we may experience.

15. Unlikeable or deeply flawed protagonists. I can't enjoy a novel unless the main protagonist is basically admirable. He or she will have faults, of course, and will make mistakes; but on the whole it must be someone I might like to know, and with whose feelings I can identify. Most people, and presumably most aliens, are fundamentally good. They do the best they can in the circumstances in which they are placed. To suggest otherwise by featuring exceptions distorts readers' view of the future.

Obviously, if I don't enjoy fiction with any of the above themes my "to be read" list is rather short. Normally I read mostly nonfiction. But if anyone has suggestions for novels I might like, I welcome them.

.

About the Author

Sylvia Engdahl is the author of eleven science fiction novels. Six of them are Young Adult books that are also enjoyed by adults, all of which were originally published by Atheneum and have been republished, in both hardcover and paperback, by different publishers in the twenty-first century. The one for which she is best known, *Enchantress from the Stars* was a Newbery Honor book in 1971, winner of the 1990 Phoenix Award of the Children's Literature Association, and a finalist for the 2002 Book Sense Book of the Year in the Rediscovery category. Her trilogy *Children of the Star* was reissued in a single volume as adult science fiction.

Engdahl's five most recent novels, a duology and a trilogy, are not YA books and are not appropriate for middle-school readers, but will be enjoyed by the many adult fans of her work. In addition, she has issued an updated and expanded edition of her nonfiction book *The Planet-Girded Suns: Our Forebears' Firm Belief in Inhabited Exoplanets* (first published by Atheneum in 1974 with a different subtitle) as well as three ebooks of collected essays.

Between 1957 and 1967 Engdahl was a computer programmer and Computer Systems Specialist for the SAGE Air Defense System. Most recently she has worked as a freelance editor of nonfiction anthologies for high schools. Now retired, she lives in Eugene, Oregon, and welcomes visitors to her website www.sylviaengdahl.com, which contains many of her essays, including those dealing with her long-term advocacy of space colonization.

CURRENTLY AVAILABLE EDITIONS OF
SYLVIA ENGDAHL'S BOOKS

Click on the title to see the book description, reviews, excerpts, and purchase links. All are available in inexpensive ebook editions; titles marked pb and/or ab also have paperback and/or audiobook editions

YOUNG ADULT NOVELS
Enchantress from the Stars - pb, ab
The Far Side of Evil - pb, ab
Journey Between Worlds - pb, ab

CHILDREN OF THE STAR
(Trilogy - YA, reissued as adult)
This Star Shall Abide (Book 1, aka *Heritage of the Star)* - pb, ab
Beyond the Tomorrow Mountains (Book 2) - ab
The Doors of the Universe (Book 3) - ab
Children of the Star: The Complete Trilogy (Omnibus) - pb

THE FOUNDERS OF MACLAIRN
(Duology - adult)
Stewards of the Flame (Book 1) - pb, ab
Promise of the Flame (Book 2) - pb, ab

THE CAPTAIN OF *ESTEL*
(Trilogy - adult)
Defender of the Flame (Book 1, aka *Passage to Destiny*) - pb, ab
Herald of the Flame (Book 2, aka *A Ship Named Hope*) - pb, ab
Envoy of the Flame (Book 3, aka *Mission to Earth*) - pb, ab
The Captain of Estel: The Complete Trilogy (Omnibus)

YA ANTHOLOGY
(Editor)
Anywhere, Anywhen: Stories of Tomorrow

NONFICTION
*The Planet-Girded Suns: Our Forebears' Firm Belief in Inhabited
    Exoplanets* - pb

COLLECTED ESSAYS
From This Green Earth: Essays on Looking Outward - pb, ab
Selected Essays on Enchantress from the Stars and More: A
    Sampler - pb, ab
Reflections on Enchantress from the Stars *and Other Essays*
The Future of Being Human and Other Essays

www.ingramcontent.com/pod-product-compliance
Lightning Source LLC
Chambersburg PA
CBHW021205020426
42331CB00003B/206